194

way to tackle technical history! I hope you like this book as much as I do.

Love always,
Pat

NOW YOU'RE LOGGING

NOW YOU'RE LOGGING
By
BUS GRIFFITHS
with a foreword by
JACK HODGINS
BUS Griffiths

First trade paper edition 1990

Harbour Publishing
P.O. Box 219
Madeira Park, BC
Canada V0N 2H0

Cover design by Roger Handling
Printed and bound in Canada

Canadian Cataloguing in Publication Data

Griffiths, Bus, 1913–
Now you're logging

ISBN 1-55017-033-3

1. Logging – Caricatures and cartoons.
I. Title.
PN6790.C2G75 1976 C813'.54 C79-001687-7

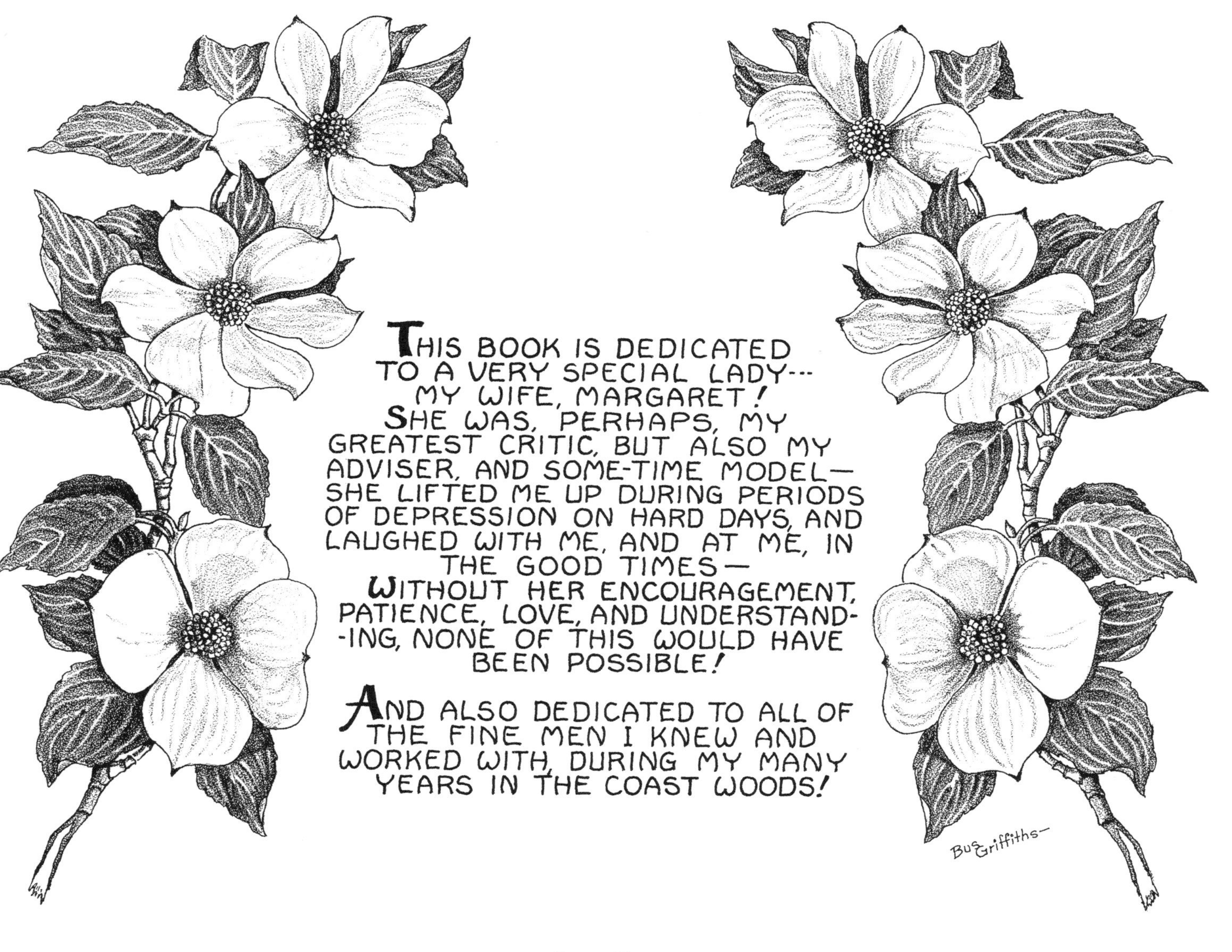

THIS BOOK IS DEDICATED
TO A VERY SPECIAL LADY---
MY WIFE, MARGARET!
SHE WAS, PERHAPS, MY
GREATEST CRITIC, BUT ALSO MY
ADVISER, AND SOME-TIME MODEL—
SHE LIFTED ME UP DURING PERIODS
OF DEPRESSION ON HARD DAYS, AND
LAUGHED WITH ME, AND AT ME, IN
THE GOOD TIMES—
WITHOUT HER ENCOURAGEMENT,
PATIENCE, LOVE, AND UNDERSTAND-
-ING, NONE OF THIS WOULD HAVE
BEEN POSSIBLE!

AND ALSO DEDICATED TO ALL OF
THE FINE MEN I KNEW AND
WORKED WITH, DURING MY MANY
YEARS IN THE COAST WOODS!

Foreword

Legends seem to grow quite naturally in Fanny Bay. To the tourist rushing up the Island highway this may seem like little more than a few miles of picturesque coastline, with the Fanny Bay Inn (Licensed Premises) and a beached ship converted to seafood restaurant brightly painted up for local colour, but for the person who stops for any length of time this community quickly comes alive with stories about the adventures of unique residents, present and past. They will tell you here of the famous bounty hunter Cougar Smith who came in to poison wolves in the early days of the settlement and whose reward has been to have his name attached to both a nearby creek and a new subdivision of water-view lots. They will tell you of the train brakeman who used to entertain the children on a flatcar Sunday mornings with fairy tales that all began with "Once upon a time there was this man who lived in Fanny Bay..." and went on to pin down faults and put everyone in his proper place. They'll tell you of the local logger who was never without a wad of "snoose" in his mouth, even while he slept, and allowed his bottom lip and stomach to be badly damaged rather than spit the juice like everyone else. They may also tell you of the fellow in that house across from the cedar swamp who worked for years in the logging camps and then made a name for himself painting pictures and writing comic strips and publishing a big novel that'll show you everything there is to know about logging around these parts in the dirty thirties.

In an old photo from those times, he stands in west coast forest with his strong hands tucked behind his regulation braces and into the waist-band of his pants, his head tilted up like a modern Paul Bunyan looking over the world he's about to destroy. In real life Bus Griffiths is less forbidding. Those hands are still as powerful as ever, the barrel chest is still impressive, but this face is not the face of either giant or destroyer. If Griffiths is destined to enter local legend beside Cougar Smith and all those others, then he will be taking his shy grin, his shining little-boy eyes, his slow drawly speech, and his gentle considerate manner right along with him.

Bus Griffiths is retired now. From logging and from commercial fishing, that is, but not from anything else. Life, if anything, is busier now than ever. Radio phone-in shows. Interviews. Strangers at the door. The subject of all this attention is more than a little bewildered. He wonders what he has done. All he intended to do was combine what he likes best with what he does best: logging and story-telling and drawing.

What he has accomplished, however, is considerable. His paintings hang in private homes around the district. Eight of his oils, depicting scenes of logging, hang in the nearby Courtenay District Museum. Two of his murals hang in the Fanny Bay Community Hall. His comic strips have been published in *The B.C. Lumberman* and distributed at Truck Loggers conventions as well as at the International Forest Equipment Exhibition in Vancouver. Some black and white drawings have been published with Peter Trower's poetry in a book entitled *Bush Poems*, and six of these – a series depicting the preparation of a spar tree – hang on the walls of a Vancouver pub called The Molly Hogan.

The primary cause of most of this fuss, however, has been the publication of his novel in comic book form, *Now You're Logging*. The term "comic book" is misleading, though, because there's no question that Bus Griffiths wants this book to be taken seriously. This is a very special and innovative kind of history.

Comic book novel, history, whatever it is, people are reading it. Loggers and ex-loggers are reading it because they can see themselves and their jobs depicted in it with humour and care. The sons and daughters of loggers are reading in it about the heroes they only heard about as kids. People who have no connection with logging are reading it because they can see it offers not only a simple and fascinating account of the way things once were in this part of the world but also an entertaining story with likable characters. Maybe Bus Griffiths thought he was simply doing what he likes to do but there's not question that he's started a bit of a fuss that won't die down for a while.

As a child in Vancouver, Griffiths dreamed of being a cartoonist. He drew comic strips and, naturally, expected them to be published. Neither of the local papers was interested, however, but Griffiths visited so often and hung around so long that the resident cartoonist of the *Vancouver Sun* eventually took the time to give him a few tips. "That was awful crude stuff I was packing around," he says now, and eventually the managing editor of *The Province* (who felt the same way, apparently) lost patience with him and told him to "go away and take some drawing lessons, kid."

Griffiths never did take that editor's advice. In fact, artistic ambitions were temporarily shelved. He had already embarked on another career. At the age of twelve he was given his first logging "job" by his father, who bought him a saw and axe and commissioned him to fall and buck the backyard alders for firewood. A few years later he was hired by a neighbour to cut cord wood for two dollars a week and two meals a day. Impatient with the length of time the job was taking, he created a silent partner for the other end of the saw out of a sapling and an inner tube. This innovation, called "falling with a chinaman" (probably an implied acknowledgement of the racist wage scales of the day) was to establish a pattern of creative approaches to the job which would continue all his working life.

Though he took his training in office work – even by the age of fifteen he was typing and bookkeeping in an office – he soon discovered that he "admired the husky guys in the warehouse over the office men" and when a boss told him he was not to fraternize with those common labourers he made what seemed an obvious choice. An office was no place for a man like him. Soon he was working in logging camps all over the Fraser Valley and up the B.C. coast.

"There's nothing I'd rather do than log," he says. And for much of his life that is exactly what he did, mostly in truck logging shows, the smaller camps, where he worked at nearly every job that kind of life had to offer – rigging, chasing, loading, booming, roadbuilding, falling, and bucking. Everything but donkey engineer – he was never much of a mechanic, he admits. In 1944 he moved to Fanny Bay on Vancouver Island with his new wife Margaret, the daughter of an English bootmaker who'd retired to a farm in Port Coquitlam and fell timber for the nearby Beban Logging Company. Only when the job was threatening to take him away from home and his growing family for long periods of time did he quit to take up commercial fishing.

In spite of his love for the logger's life, Griffiths was never very proud of what the industry was doing to the landscape. The ruined streams, the displaced wildlife, the wasted trees left to rot—all made him uncomfortable about his role in this whole operation. "Years later I would go back hunting to some spot where I'd worked—after the slash had been burned and the fireweed grown up—and I'd find the streams' banks three times deeper than before, the water red with mud, everything eroded; I never felt very proud of that."

And yet there is no question of Griffiths' feelings for the men who caused this damage. He recalls with obvious fondness the high-rigger who always sat on the top of a spar tree just long enough to roll a smoke, then dropped his hat and beat it to the ground. In *Now You're Logging*, a similar character teaching the protagonist how to top a tree likes to sit on the top two hundred feet in the air, because, he says, it helps him to unwind. "When you're up here you're on your own—you don't have to look out for a bunch of other bastards." When he races his hat to the ground, he drops in great dizzying spirals, ten feet or more at a leap.

This may seem to be the stuff of legend, but there are many such larger than-life people in Griffiths' repertoire. The tough-talking camp-foreman named Donnegan in the novel is remembered as a two hundred and twenty-five pound brute in real life, a man who terrorized the camp with his loud voice and impressive strength. A whistle punk who went to sleep on the job soon learned to keep awake when "Donnegan" rushed like a mad bull through the brush and leapt right on top of him. Nothing mattered but getting the logs out; suffering from a severe attack of arthritis, he refused to sleep in his bunk in case he wasn't able to get out of it in the morning and would have to miss a day's work. Instead, he slept in a chair, slumped over a table. Pesky flies were no safer than whistle punks in this man's show; he could spend his lunch-break pulling their wings off and watching them crawl. And yet, Griffiths recalls seeing this same man hold up the operation long enough for him to rescue a toad from a dangerous spot under a log and move it to a safer place in a hollow tree. Like most humans, he was full of contradictions.

And then there was the fellow who threw temper tantrums. Enraged, he whipped off his hat, spat his snoose into it, dropped the hat to the ground where he jumped up and down on it cursing while his hands pulled his hair down over his face. Eventually, when he'd worked off his anger, he picked up the hat and wiped it out with his elbow, then looked around at his audience. "Alright boys," he said in a surprisingly gentle voice, "let's get to work."

Not all of Griffiths' colourful loggers got into the novel, however, and those who did were changed a little, or combined, as in most fiction. Getting the feel of the characters is an important part of the story-telling process as far as he is concerned.

Bookshelves in the Griffiths home display some of the titles you might expect to find there. *Sometimes a Great Notion*, Ken Kesey's novel of an Oregon logging family, is there, as well as E.G. Perrault's *The Kingdom Carver* a novel recording the rise of a Vancouver Island timber baron. Though Roderick Haig-Brown's logging novel *Timber* is not there, Griffiths recalls reading it, and indeed recalls reading most other Haig-Brown novels to his sons. The novelists who had the greatest influence on the storyline of *Now You're Logging* are not Kesey or Haig-Brown or Perrault, but the writers of those dozens of Westerns that he has enjoyed reading over the years, Luke Short and Zane Grey. Like the classic Western, Griffiths' plot is an adventure story, where two pals are pitted against the hardships of a frontier world, and life is risked over and over in the long exciting process of initiation. Even at the end, when the hero finally "gets the girl" the reader has the feeling that—like the cowboy and his horse—Al's first love will always be the woods.

Even the friendship between Al and Red is reminiscent of the Western and other adventure novels. Having a "partner" has always been as important to the logger as to the cowboy. In that male world so far from women and family, there must be a close friend to share letters, complaints, and dreams.

All of this drawing and creating was done in the little studio Griffiths built by walling in a patio at the back of his house. Surrounded by the comfortable clutter of magazines and papers heaped on a desk, brushes and paints spread over a table, he produced each page at a tilted draftsman's table using grease pencils for the drawings and a draftsman's technical pen for the dialogue and the footnotes of logging terms.

When he isn't painting or creating cartoon strips or signing books, Griffiths spends a good deal of time out of doors. In his backyard he cultivates a garden much larger than he and his wife can use. Out on the Strait he fishes from his little outboard. And back in the woods he hunts for a winter's supply of venison. He also lifts weights regularly, a pastime he adopted when he gave up logging for fishing and discovered he'd "started to go soft."

The dog, the cat, the rabbits, make travel difficult, but the Griffiths do manage to see something of their sons once in a while. Both are objects of obvious parental pride. Bert, a Staff-Sergeant in the RCMP, has worked in the Ottawa computer division; and Steve, a scholarship student while at university, has been a marine biologist working for Ontario Hydro. There is no doubt that the energy, pride and enthusiasm of Margaret Griffiths have had much to do with the success achieved by sons and husband alike. The dedication to Margaret Griffiths at the front of *Now You're Logging* is no mere sentimental formality. Much more than this one book is implied. There is a strong sense that the Griffiths, husband and wife, are a team; they're in all of this together.

And what does the future hold for Bus Griffiths now, aside from occasional media attention and a sort of immortality on museum walls? He'll continue to hunt and fish and work in his garden, he'll do more illustrating, he'll produce more oil paintings of logging scenes. He'll go on getting better at his work. "If you like what you're doing," he says, "I don't think you can do anything but improve."

Some day, when the people of Fanny Bay tell you stories of Cougar Smith who poisoned wolves, or the train brakeman who told fairy tales with a local cast, or the snoose-chewer who refused to spit, they may also tell you of the day a few years back when Bus Griffiths, the painter and book-writer, tore his hand open with a power saw. Rushing to the hospital for stitches to the bleeding wound, he took time to stop and help someone who was having trouble changing a flat tire at the side of the road. And when he got home with his new stitches and white bandages he went back out into the bush immediately to finish the interrupted job. Nothing special about that, nothing to brag of there—it was nothing more than what that gruff arthritic foreman would have expected of a man like Griffiths. A perfectly natural way to act for any old logger worth his salt. Like Grandaddy Tough in Peter Trower's poem, Bus Griffiths "has walked with legends / and all unknowing / become one."

JACK HODGINS

NOW YOU'RE LOGGING

A STORY BASED ON LOGGING IN THE "HUNGRY YEARS," AND ABOUT THOSE COMPLEX, AND INTRIGUING CHARACTERS, THE LOGGERS, PRODUCTS OF A SPECIAL TYPE OF WORK, AND WAY OF LIFE---- HARD, TOUGH MEN, BECAUSE THEIR LIFE WAS TOUGH, AND BECAUSE OF THE VERY NATURE, AND THE DANGER OF THEIR WORK—

THEY LIVED, AND QUITE OFTEN DIED, WITH THE WIND AND THE RAIN IN THEIR FACES, AND THE PUNGENT ODOR OF THE BIG WOODS IN THEIR NOSTRILS—

AND OCCASIONALLY, FOR NO APPARENT REASON, THEY WOULD "PULL THE PIN," AND UNWIND WITH A HELL-ROARING SPREE—

TRUE, THEY WERE A HARD-LIVING, HARD-DRIVING BUNCH······AND SURE, THEY WERE QUITE CRUDE AT TIMES·····BUT EVEN AMONG THE CRUDEST, MEANEST, AND THE TOUGHEST, THERE WAS A GENTLER SIDE, AND A GREAT LOVE OF NATURE AND THE OUTDOORS—

THESE WERE THE LOGGERS, MEN OF STEAM AND GAS DONKEYS, HIGH-LEAD LOGGING, AND THE BIG TIMBER OF THE RAIN FORESTS OF THE FOG-SHROUDED PACIFIC COAST······MEN WHO TOOK A PRIDE IN THEIR WORK AND SKILLS, AND WHO WERE ARTISTS IN THEIR OWN RIGHT·····

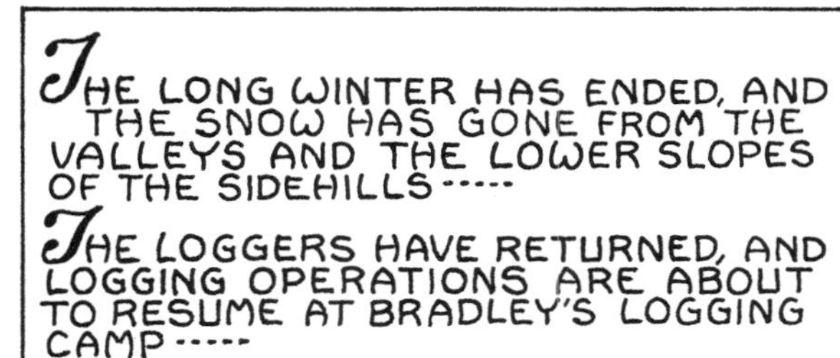

LOGGING TERMS

*"PUSH"—CAMP FOREMAN—
*"HOOKING"—HOOKTENDER- HE IS IN CHARGE OF THE WHOLE OPERATION OF GETTING LOGS FROM WOODS TO LANDING AT THE SPAR TREE—
*"TRACKSIDE"—A LOGGING SETTING WHERE THE SPAR TREE IS LOCATED AT THE ROAD, OR TRACKSIDE—
*"RIGGING SLINGER"— IN CHARGE OF THE CHOKER-MEN, AND RESPONSIBLE FOR THE OUTPUT OF LOGS
*"BY THE FACE"— LOGGING TERM MEANING "BEING IN CHARGE"—
*"CHUNK OUT"— MEANS TO CLEAN OUT--- GET RID OF SOMETHING—

LOGGING TERMS

*"CRUMMY"—THE TRUCK OR RAILROAD CAR USED TO HAUL THE CREW TO AND FROM WORK—

*"CATTY"—THE ULTIMATE COMPLIMENT FOR A RIGGING MAN—MEANS QUICK ON HIS FEET AND GOOD ON THE JOB

AL SHOUTS THE "GO AHEAD" SIGNAL TO THE WHISTLE PUNK, AND ONE SHRILL BLAST OF THE STEAM WHISTLE PIERCES THE CLEAR MORNING AIR— THE ROAR OF THE BIG YARDING DONKEY FLOATS UP OVER THE SIDEHILL AND TWO LOGS POP OUT OF THE BRUSH ON THE ENDS OF THE TWO STEEL CHOKERS —THEN AL CALLS TO THE PUNK—

WE ARE NOW READY TO STRING OUT OUR NEW "YARDING ROAD"—THIS IS DONE BY THE HOOKTENDER, WHO CONTROLS THE POSITIONING OF THE LINES ······
A LIGHT LINE CALLED *"STRAWLINE" IS USED FOR CHANGING ROADS—THE HOOKER PULLS THE LINE OUT OVER THE NEXT STRIP OF FELLED AND BUCKED TIMBER HE WISHES TO REACH WITH HIS YARDING RIGGING ···

LOGGING TERMS

*"TURN"— THE NUMBER OF LOGS YARDED IN ONE TRIP OF THE CHOKERS

*"ROAD"— THE LINE FROM SPAR TREE TO BLOCK AT BACK END, ON WHICH LOGS ARE HAULED, IS CALLED a "ROAD"— AS EACH STRIP IS LOGGED OFF THE ROAD IS CHANGED SO THAT MORE LOGS CAN BE REACHED—

BUS Griffiths—

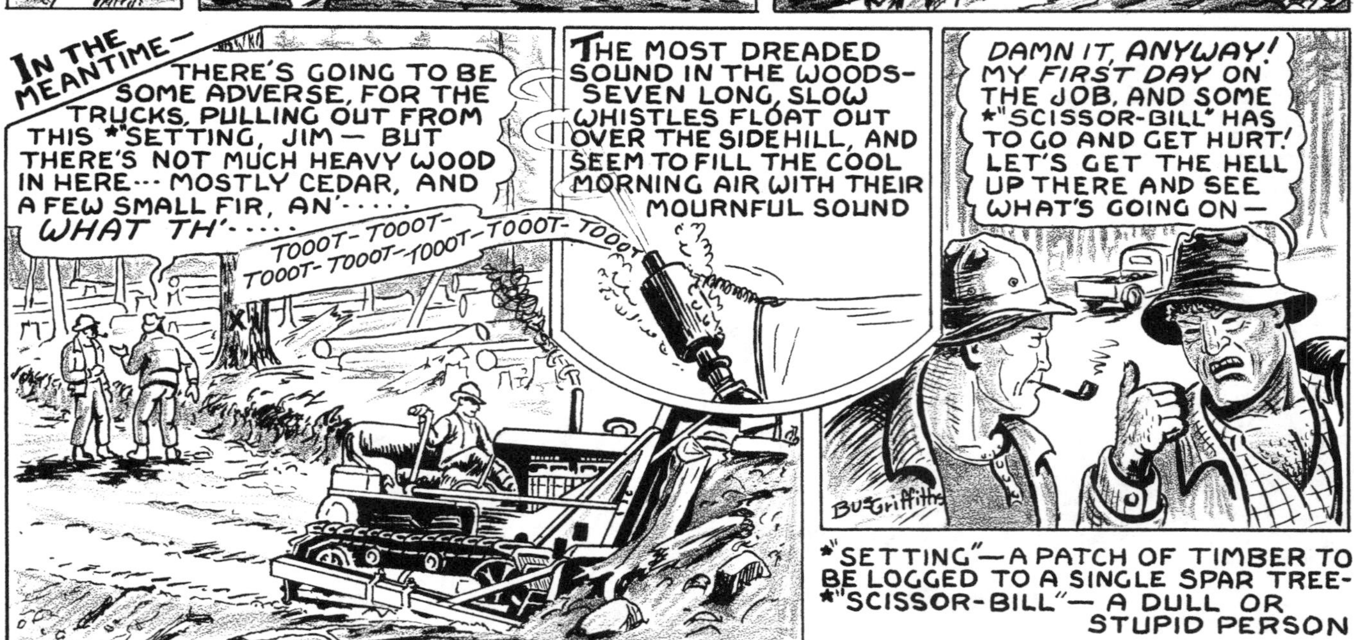

*"SETTING"—A PATCH OF TIMBER TO BE LOGGED TO A SINGLE SPAR TREE-
*"SCISSOR-BILL"— A DULL OR STUPID PERSON

SURE HOPE OL' HENDY'S OKAY!
GOD! THAT RIVER SURE SOUNDS A LONG WAY DOWN!
WELL, EVERYTHING'S ALL SET, BUT I'M ALMOST SCARED TO GO DOWN···· GUESS I'M AFRAID OF WHAT I MIGHT FIND DOWN THERE!

I'LL SIGNAL WHEN I GET DOWN TO HENDY—THEN YOU PULL UP AND SEND DOWN THE STRETCHER! BETTER SEND DOWN SOME OF THOSE SPLINTS I CUT, TOO!
SURE, AL!

IF HENDY HAD TO FALL INTO THE CANYON HE COULDN'T HAVE PICKED A BETTER PLACE!··· IT'S NOT NEAR AS STEEP HERE AS IT IS FURTHER ALONG!

HE WENT THRU THIS CLUMP OF BRUSH! SURE LOTS OF BROKEN BRANCHES···· BUT THAT WOULD HELP SLOW HIM DOWN! HE MAY NOT BE HURT TOO BAD!!
THERE HE IS! ON THAT LEDGE!!!

IF HE HADN'T STOPPED HERE HE'D HAVE GONE CLEAR TO THE RIVER!

HENDY! CAN YOU HEAR ME ?

HE'S GOT A BAD CUT ON HIS HEAD, AND HE'S SURE SCRATCHED UP···· AH! HE'S STARTING TO COME OUT OF IT!!

THE CREW PULLS THE INJURED HENDY TO THE TOP OF THE CANYON JUST AS ART DONNEGAN, THE NEW PUSH, AND BOSS-LOGGER, JIM BRADLEY ARRIVE ON THE SCENE ~

CLEAN THE SNOW OFF THAT GODDAMNED STUMP, AND SET HIM DOWN!
1

BOY! THAT'S REALLY LAYIN' DOWN ON THE JOB— HOW IN HELL DO YOU EXPECT TO GET LOGS LAYIN' ON THE FLAT OF YOUR BACK ON THAT STRETCHER?.... YOU HURT BAD, BOY?
NOPE—JUST RESTIN', ART! I'VE BEEN WORKIN' TOO HARD!
2

YOU SURE LOOK LIKE HELL! ARE Y'SURE YOU AIN'T HURT BAD, JIM?
I FEEL KINDA BEAT UP..... AN' I THINK I'VE GOT A BUSTED LEG!
3

DAMN IT! I'M SURE SORRY, JIM! HOW DID IT HAPPEN, ANYHOW?
I COME UP TO SEE HOW MANY LOGS THERE WAS UP HERE, AN' HOW BAD THE SNOW AN' ICE WAS—I SLIPPED AN' WENT OVER THE EDGE!
4

BUT ART— YOU SHOULD HAVE SEEN THAT AL GUY PERFORM! THAT GUY'S REALLY GOOD! HE SEEMED TO KNOW ALL THE RIGHT MOVES—HE PUT THE SPLINTS ON MY LEG, TOO, AN' IT DON'T HURT SO MUCH, NOW!
HE'S GOOD, EH? WELL, I WANT 'EM GOOD AND TOUGH! I'M GONNA FIND OUT IF HE'S GOT ANY SAND IN HIS CRAW!
5

IN THE MEANTIME, AL IS BEING PULLED UP FROM THE LEDGE WHERE HE HAD FOUND THE INJURED HOOKTENDER, JIM HENDY
6

BUS GRIFFITHS
7

AL REACHES THE RIM OF THE CANYON IN TIME TO HEAR THE FOLLOWING —
OKAY! THE CRUMMY IS WAITIN' AT THE LANDING AN' THE SOONER HENDY GETS TO THE SAWBONES THE BETTER... C'MON MOVE IT!
I NEVER THOUGHT IT WOULD BE HENDY WHO WAS HURT! I THOUGHT IT WOULD BE ONE OF YOU OTHER *"RANCHERS"!

THE FIRST DAY BACK LOGGIN' AN' I FIND YOU BASTARDS PLAYIN' AROUND HERE LIKE A BUNCH OF GODDAMNED SWISS MOUNTAIN CLIMBERS! WE GOT LOGS TO TAKE OUT!

LISTEN, MISTER! WE WASN'T PLAYING AROUND! A MAN WAS BADLY HURT---- COULD EASY HAVE BEEN KILLED! AND AFTER THAT SESSION IN THE CANYON I DON'T FANCY HEARIN' A BUNCH OF GUFF! I'VE GOT A GOOD NOTION TO *"BUNCH IT"!
WHOAH! DON'T GET EXCITED, AN' DON'T "BUNCH IT"! I DON'T WANT TO SEE ANYONE HURT, EITHER- -BUT I DO WANT TO GET LOGS----AN' TH' SOONER YOU GUYS FLY AT IT, TH' BETTER

TOUCHY... AIN'T HE !?!
WELL, WHAT DO YOU EXPECT?...YOU SHOULDN'T TRY TO BE SO TOUGH, ART! YOU SHOULDN'T PUSH 'EM SO HARD!

WE'LL HAVE TO SEND TO TOWN FOR A NEW HOOKER, NOW THAT HENDY IS HURT!
LIKE HELL WE WILL! AL CAN HANDLE IT! TH' GUY'S GOOD AN' HE CATCHES ON FAST! HENDY AN' TH' BOYS SAID HE TOOK CHARGE OUT HERE AN' HE KNEW ALL TH' RIGHT MOVES!
YOU KNOW, JIM, A GUY'S GOT TO BE A BIT OF A HARDROCK TO BOSS A CREW— THAT'S WHY I SPOKE TH' WAY I DID----I WANTED TO SEE AL'S REACTION!
HE'LL DO! HE'S TOUGH AN' HE CAN HANDLE TH' CREW!
DID YOU SEE HOW HE HAD IT RIGGED TO GET HENDY OUT?... AN' TH' SNUBBIN' ROPE TO KEEP TH' STRETCHER AWAY FROM TH' CANYON WALL?.... YEAH....AN' WE'LL PUT YOUNG RED PULLIN' RIGGIN'—I THINK THEY'LL MAKE A GOOD TEAM...THEN MAYBE WE'LL GET LOGS!
WELL! I'LL BE DAMNED!
Bus Griffiths
LOGGING TERMS — *"RANCHERS"— GREENHORNS — *"BUNCH IT"— TO QUIT

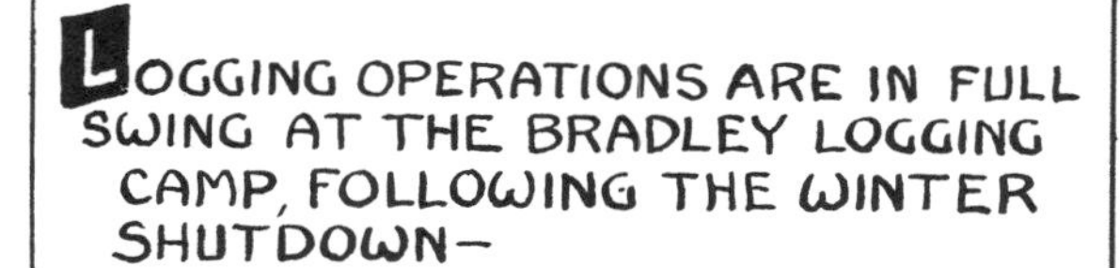
LOGGING OPERATIONS ARE IN FULL SWING AT THE BRADLEY LOGGING CAMP, FOLLOWING THE WINTER SHUTDOWN—

WELL, ART— THAT WAS A SMART MOVE WHEN YOU PUT AL TENDIN' HOOK— WE'VE SURE BEEN GETTIN' OUT LOTS OF LOGS SINCE!
YEAH— HE'S DOIN' FINE--- AN' WITH RED PULLIN' RIGGIN' THEY MAKE A GOOD TEAM!

THEY'VE GOT THE REST OF THE GUYS WORKIN' GOOD, TOO— THEY SURE ARE A *"LOG-HUNGRY" CREW!
YEAH— THEY'VE GOT THE LANDIN' PLUGGED WITH LOGS MOST OF THE TIME!

MAYBE THE TRUCKS CAN GET A FEW EXTRA LOADS----- OR MAYBE "CHILLIWACK" CAN THROW A BIT MORE WOOD ON 'EM!
LIKE HELL! ON BOTH COUNTS

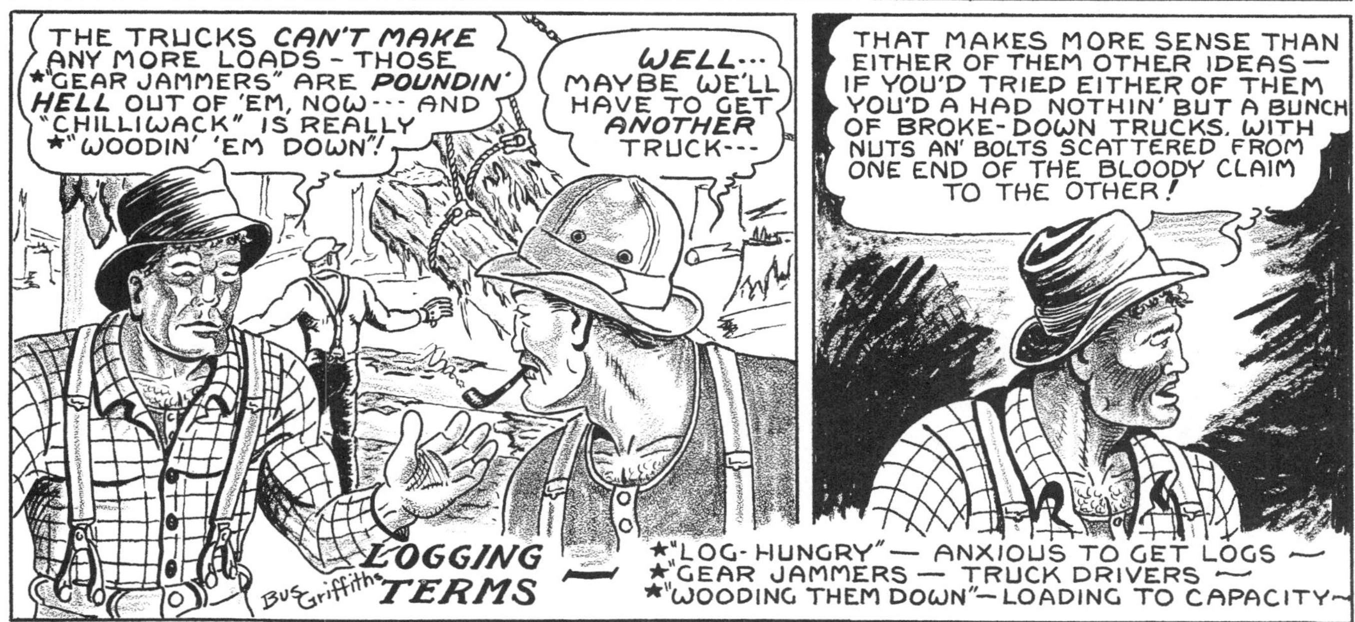
THE TRUCKS CAN'T MAKE ANY MORE LOADS - THOSE *"GEAR JAMMERS" ARE POUNDIN' HELL OUT OF 'EM, NOW--- AND "CHILLIWACK" IS REALLY *"WOODIN' 'EM DOWN"!
WELL... MAYBE WE'LL HAVE TO GET ANOTHER TRUCK---
THAT MAKES MORE SENSE THAN EITHER OF THEM OTHER IDEAS— IF YOU'D TRIED EITHER OF THEM YOU'D A HAD NOTHIN' BUT A BUNCH OF BROKE-DOWN TRUCKS, WITH NUTS AN' BOLTS SCATTERED FROM ONE END OF THE BLOODY CLAIM TO THE OTHER!
Bus Griffiths
LOGGING TERMS —
*"LOG-HUNGRY"— ANXIOUS TO GET LOGS —
*"GEAR JAMMERS — TRUCK DRIVERS —
*"WOODING THEM DOWN"— LOADING TO CAPACITY—

LOGGING TERMS

*"TOPPING AND RIGGING"- WHEN A TREE WAS SELECTED AS A SPAR, THE LIMBS AND TOP HAD TO BE REMOVED- THE TREE WAS THEN RIGGED WITH GUYLINES AND PULLEY-TYPE BLOCKS, THRU WHICH THE YARDING LINES FROM THE DONKEY WERE THREADED — THIS JOB WAS DONE BY A MAN CALLED THE "RIGGER", OR "HIGH RIGGER", IN A METHOD OF LOGGING KNOWN AS THE HIGH LEAD SYSTEM ~

Bus Griffiths

AL, GET YOUR RIGGIN' SLINGER TO TAKE OVER FOR THE REST OF THE AFTERNOON---AND YOU COME WITH ME!
OKAY, ART! I'LL SEE RED---- BE RIGHT WITH YOU
LATER
C'MON— WE'LL TAKE A RIDE UP TO THE NEXT SETTIN'----YOU CAN'T HAVE MANY LOGS LEFT ON THIS ONE!
NOPE— WE SHOULD BE FINISHED LOGGING AND READY TO MOVE IN TWO DAYS!
GOOD! THAT'LL WORK OUT FINE---- SAY, AL-- Y'KNOW, THIS OUTFIT CAN'T AFFORD TO PAY A MAN JUST TO TEND HOOK!
THIS IS ONE OF THEM "HOOK AND RIG OUTFITS"---THEY EXPECT THE HOOKER TO BE ABLE TO CLIMB, AND DO HIS OWN TOPPIN' AND RIGGIN', TOO!
I'VE BEEN EXPECTING THAT— I'VE DONE A BIT OF CLIMBING WITH THE BELT AND SPURS— BUT I'VE NEVER TOPPED A TREE, OR DONE ANY RIGGING—
— BUT I'D SURE LIKE TO LEARN !!
GOOD! AN' THERE'S NO TIME LIKE THE PRESENT! WE'VE GOT A REAL SNARLY OL' FIR FOR YOU TO START ON!
THERE SHE IS!
WOW! IT'S SURE A ROUGH LOOKING TREE FOR A SPAR!
YEAH— BUT SHE'S SOUND AN' SHE SHOULD MAKE A GOOD TOUGH ONE--- BESIDES— IF WE USE THIS TREE, IT'LL SAVE US HAVING TO ★"RAISE A SPAR"
NOW, LOOK--- THIS IS YOUR OUTFIT FROM NOW ON---- SO TAKE CARE OF IT— YOUR LIFE COULD DEPEND ON IT!
OKAY— THANKS!
BUS GRIFFITHS
LOGGING TERMS ★"RAISE A SPAR"— A SUITABLE STANDING TREE COULD NOT ALWAYS BE FOUND NEAR THE SPOT WHERE A SPAR WAS NEEDED---THEN ONE WOULD BE BROUGHT IN FROM ELSEWHERE, AND RAISED FOR THE PURPOSE—

THERE'S A SPLICED EYE IN ONE END OF THE ROPE, AND WE USE A "CAT'S PAW" HITCH TO TIE THE OTHER END TO THE EYE---LIKE THIS!---IT'S FAST AND SAFE, AND IT'S EASY TO ADJUST THE SLACK IN YOUR ROPE—
WHEN YOU'RE CLIMBIN' DON'T TRY TO HUG THE TREE---IT'S NOT A GIRL! GIVE YOURSELF ENOUGH SLACK AN' LEAN BACK ON YOUR ROPE— THIS ROPE HAS A STEEL CORE AN' YOU DON'T HAVE TO WORRY ABOUT THE SONOFABITCH BREAKING!
ONE BIT ON YOUR AXE IS STUBBED OFF FOR CHOPPIN' KNOTS, THE OTHER IS FOR CHOPPIN' OFF THE TOP, AND IT'S RAZOR SHARP— IT CAN CUT THRU THAT ROPE WITH ONE SWIPE---SO DON'T YOU FORGET----
NOW, WE'RE GONNA CLIMB UP AN' KNOCK OFF THE LIMBS, AN' THEN WE'LL TOP HER! JUST TAKE YOUR TIME AN' DON'T GET EXCITED!
Bus Griffiths
WELL, LET'S GO! WE'LL CLIMB UP TOGETHER—
JUST TAKE YOUR TIME—YOU'RE DOIN' FINE!

I'M GLAD THIS AIN'T A SPRUCE— WITH BIG LIMBS LIKE THIS, WE'D HAVE TO USE A SAW TO CUT 'EM---THOSE BIG SPRUCE LIMBS ARE SURE *HELL* ON AN AXE!
DID YOU USE A SAW MUCH, WHEN YOU WERE TOPPIN' SPARS?

NOPE— YOU GOTTA HAVE THE AXE, ANYWAY, TO CHOP LIMBS, AN' WHEN YOU'RE GOOD WITH AN AXE IT DON'T TAKE LONG TO KNOCK OFF THE TOP! BESIDES, IT'S *AWFUL EASY* TO GET CUT WITH THEM DAMN SAWS—

WHEN YOU'RE CHOPPIN' LIMBS THE SAW IS HANGIN' FROM YOUR BELT BY A ROPE— SOMETIMES A FALLIN' LIMB WILL HIT THE SAW AN' FLIP IT RIGHT UP AT YOU—I REMEMBER WHEN I WAS A KID UP AT THE C.R.T. CAMP AN' A RIGGER GOT CUT THAT WAY----

3

HE NEAR BLED TO DEATH BEFORE HE GOT DOWN OUT OF THE TREE— I WAS SURE SCARED, AN' SWORE I'D NEVER GO TOPPIN' TREES-- BUT, *HERE I AM!*
YEAH—A GUY NEVER KNOWS--- I NEVER THOUGHT I'D EVER BE WORKING THIS HIGH ABOVE THE GROUND!
4

LATER
THIS IS HIGH ENOUGH---YOU CAN KNOCK OFF THE TOP HERE
5

NOW LOOK—THIS IS HOW YOU MAKE THE SIDE NOTCHES- YOU CUT THRU THE SAPWOOD AN' INTO THE HEARTWOOD—
6

A SPLITTIN' TREE IS ONE OF THE BIGGEST DANGERS WHEN YOU'RE TOPPIN'---BUT IF YOU PUT YOUR SIDE NOTCHES IN *RIGHT,* THERE AIN'T MUCH DANGER--- THE TOP WILL BREAK CLEAN!
Bus Griffiths
7

THE SIDE NOTCHES AND UNDERCUT ARE OKAY— YOU CAN MAKE YOUR BACK CUT NOW!
YEAH-SURE
1

LATER
HOLD IT A MINUTE, AL!
2

SHE'S JUST ABOUT READY TO GO----KNOW WHAT YOU SHOULD DO NEXT?
PRAY?
3

HAW! HAW! HAW! WELL----I GUESS YOU COULD DO THAT, TOO-- BUT THAT AIN'T WHAT I MEANT!
4

5
WHEN YOU'RE CHOPPIN', THE JOLT OF THE AXE DRIVES YOUR SPURS DEEPER AN' DEEPER INTO THE TREE— YOU HAVE TO TWIST AN' WIGGLE YOUR FEET TO GET THE SPURS FREE----AN' THEN JAB 'EM IN, BUT NOT TOO HARD!

IF YOU DIDN'T DO THAT, AND THE TREE STARTED TO SPLIT, YOU'D NEVER BE ABLE TO MOVE FAST ENOUGH TO GET BELOW THE SPLIT----OKAY-- LET HER GO!!
BUS Griffiths-
6

7
C-RA-C-K

AL FELT THE BIG TREE SHUDDER, THEN SAG OVER WITH THE PULL OF THE HEAVY TOP— THERE WAS A SHARP, SPLINTERING CRACK—
—THEN THE VIOLENT KICKBACK AS THE TOP BROKE FREE, AND FOR SEVERAL MOMENTS THE STARK, BARE TRUNK LASHED BACK AND FORTH ACROSS THE SKY, AS THOUGH IN ANGER—

AL HAS JUST HAD ONE OF THE MOST THRILLING EXPERIENCES IN THE WOODS! THE THRILL OF TOPPING HIS FIRST SPAR TREE! WITH SENSES REELING, HE STILL HAS A FEELING OF GREAT PRIDE IN HIS ACCOMPLISHMENT–

BEFORE AL REALIZES WHAT IS ABOUT TO HAPPEN, ART IS SITTING ASTRIDE THE FRESHLY TOPPED SPAR TREE, CALMLY REACHING FOR "THE MAKINGS", TO BUILD A SMOKE–

LEAN IN TOWARDS THE TREE--- I'M COMIN' DOWN— I'LL FLIP MY ROPE OVER YOU AND DROP BELOW!
OKAY, ART—

DID YOU EVER SEE A GUY BEAT HIS HAT TO THE GROUND? WELL, THAT'S "ACT TWO" IN MY LITTLE SHOW— HERE--- I'LL GIVE YOU MY AXE---
JUST WAIT UNTIL I TIE THE ROPE TO MY BELT, ART—

ART DROPS HIS HAT, KICKS HIS SPURS FREE OF THE BARK, AND STARTS DOWN THE TREE IN GREAT, DIZZYING SPIRALS, DROPPING TEN FEET OR MORE AT A LEAP—
GOOD GOD! HE'S CRAZY!!

AL WATCHED THE BREATHTAKING DISPLAY IN AMAZEMENT, AND SAW THE BURLY FOREMAN REACH THE FOOT OF THE SPAR— WITH A QUICK MOVEMENT OF HIS HANDS HE RELEASED THE "CAT'S PAW" HITCH AND FREED HIS ROPE FROM THE TREE--- THEN, STEPPING OUT, HE CAUGHT HIS FALLING HAT—

HEY, AL-- C'MON DOWN! WHAT'S TAKING YOU SO BLOODY LONG?

DON'T WORRY- I'LL BE DOWN— BUT I'M TAKING IT SLOW AND EASY!

WE MADE FAST TIME ON THE TREE, AND WE SHOULD GET TO CAMP IN TIME TO HAVE A GOOD SHOWER BEFORE CHOW-TIME!
THAT SOUNDS PRETTY GOOD TO ME! I'M ABOUT BEAT!
BUS Griffiths-

AL RICHARDS IS NOW TENDING HOOK ON THE "*TRACKSIDE," REPLACING THE INJURED HOOKTENDER, JIM HENDY— AL'S SIDEKICK, RED HARRIS, IS PULLING RIGGING FOR HIM, AND THE TWO PALS SEEM TO BE THE CATALYST THAT MOULDS THE TRACKSIDE RIGGING CREW INTO AN EFFICIENT, HIGH-PRODUCING UNIT, A "LOG-HUNGRY CREW," MUCH TO THE SATISFACTION OF BOSS-LOGGER, JIM BRADLEY, AND HIS ROUGH, TOUGH FOREMAN, ART DONNEGAN!

AL HAS BEEN INFORMED THAT IN FUTURE HE WILL HAVE TO HANDLE THE HIGH RIGGING CHORES, IN ADDITION TO HIS JOB AS HOOKTENDER— UNDER THE EXPERT GUIDANCE, AND JAUNDICED EYE OF FOREMAN DONNEGAN, HE TOPS HIS FIRST SPAR TREE, A THRILL HE WILL NEVER FORGET!

THE "OLD MAN" IS SURE IN A FOUL MOOD—HE'S GROWLIN' LIKE A BEAR WITH A SORE PAW—
OH! WHAT'S HIS PROBLEM?

HE'S MIFFED 'CAUSE THE LOADERS HAVEN'T GOT ALL THE LOGS LOADED OUT FROM THE OLD SETTIN' AN' WE CAN'T *"STRIP TH' SPAR TREE" UNTIL THEY FINISH—
WELL, WE'VE GOT THIS NEW SPAR NEARLY RIGGED, AN' WE'LL BE LOGGING TO-DAY! WHAT'S HIS BEEF?

HE WAS MAD ABOUT THE LOGS PILING UP ON THE LANDIN' AN' THOUGHT THE TRUCKS COULD HAUL AN EXTRA LOAD OR TWO, OR MAYBE "CHILLIWACK" COULD PILE MORE WOOD ON EACH LOAD

I SHOT OFF MY FACE, A BAD HABIT OF MINE, AN' TOLD HIM WHAT I THOUGHT OF THEM IDEAS-- HE'S GETTIN' MORE PRODUCTION THAN HE'S EVER HAD---BUT THESE BOSS LOGGERS ARE ALL ALIKE--- THEY SEE A BIT OF A HANGUP SOMEWHERE AN' THEY GET *"PANICKY"!

I'LL GO UP AND HANG THAT BULL BLOCK STRAP AND SHACKLE, AND THEN YOU CAN SEND UP THE BLOCK— OKAY! GOING UP!
THINK YOU CAN HANDLE IT OKAY?
IF I CAN'T I'LL HOLLER—AN' THANKS FOR HELPING ME HANG THE GUYLINES—
HELL! I HAD TO MAKE SURE IT WAS DONE RIGHT, DIDN'T I?
LOGGING TERMS
*"TRACKSIDE"—A LOGGING SETTING WITH SPAR AT ROAD OR TRACK
*"STRIP THE SPAR"— TAKE DOWN ALL THE RIGGING--BLOCKS, LINES, ETC.
*"PANICKY"— EXCITED---
BUS Griffiths—

LOGGING TERMS

*"PASSLINE"— A LIGHT STEEL CABLE USED TO TAKE THE RIGGER UP AND DOWN THE SPAR TREE, AND TO PULL UP GUYLINES, BLOCKS, ETC., WHEN RIGGING UP—

*"TAIL BLOCKS"— THE WOOD'S BLOCKS AT THE BACK END OF THE SETTING, THRU WHICH THE HAULBACK LINE RUNS, AND WHICH ALSO CONTROL THE POSITION OF THE YARDING LINES-

*"SQUARE LEAD— AT RIGHT ANGLES TO THE ROAD, OR TRACK— BECAUSE, ON A TRACK-SIDE SETTING, THE DONKEY SITS MORE OR LESS PARALLEL TO THE ROAD, OR TRACK—

Bus Griffiths

WE LEAVE THE RIGGING-UP CREW, AND JOIN RED AND HANK, WHO ARE PACKING OUT THE TAIL BLOCKS —

LOGGING TERMS

*"*MEAT ON THE HOOK, PICK UP THE SLACK*"— MEANS THAT THE RIGGER IS READY TO COME DOWN ON THE PASSLINE, AND IS A SIGNAL FOR THE ENGINEER TO TIGHTEN THE PASS-LINE AND TAKE THE RIGGER'S WEIGHT ON THE LINE—THEN THE RIGGER WILL FREE HIS ROPE FROM THE TREE AND COME DOWN ON THE LINE —

Bus Griffiths—

YOU NEEDN'T BOTHER PUTTIN' THE MACHINE *"IN LEAD"– I CHECKED HER, AN' SHE'S OKAY! YOU CAN GO AHEAD AN' THREAD THE LINES

GOOD! THAT'LL SAVE A LITTLE TIME

SHEDDING HIS BELT AND SPURS, AL TAKES THE END OF THE STRAWLINE, SITS IN THE PASS CHAIN, AND IS PULLED BACK UP THE SPAR TREE–

THE STRAWLINE IS THREADED THRU THE HIGH-LEAD, OR BULL BLOCK, FROM THE MACHINE SIDE, THEN DOWN THRU THE HAULBACK BLOCK FROM THE OPPOSITE SIDE –

AL MAKES THE END OF THE LINE FAST TO THE SHELL OF THE BLOCK, THEN PULLS SLACK ON THE BIGHT OF THE LINE BETWEEN THE BLOCKS–

AL COMES DOWN ON THE PASSLINE AND HOOKS THE END OF THE STRAWLINE TO THE EYE OF THE HAULBACK LINE, AND THE *"DONKEY-PUNCHER" ENGAGES THE FRICTION-TYPE CLUTCH ON THE STRAWLINE DRUM OF THE BIG *"DONKEY"– THE LINE STARTS TO WIND ON THE DRUM, AND THIS PULLS THE HAULBACK LINE OUT THRU THE BLOCKS AND BACK TO THE DONKEY –

Bus Griffiths–

THE STRAWLINE IS UNHOOKED, AND THE EYE OF THE HAULBACK IS FASTENED TO THE EYE OF THE MAINLINE WITH A SMALL SHACKLE THAT WILL PASS THRU THE SHEAVE OF THE BULL BLOCK--- THEN THE PUNCHER "GOES AHEAD" ON THE HAULBACK-

LOGGING TERMS

*"IN LEAD"– THE HOOK-TENDER CLIMBS BETWEEN THE DRUMS ON THE WINCH AND SIGHTS FROM THE UNDERSIDE, ALONG THE FLANGE OF THE DRUMS TO THE BLOCKS IN THE TREE, TO MAKE SURE THE LINES WILL SPOOL PROPERLY– IF THE MACHINE IS NOT IN LEAD, THE NOSE OF THE SLEIGH IS MOVED TO ONE SIDE OR THE OTHER TO MAKE THE ADJUSTMENT–

*"PUNCHER"– ENGINEER

*"DONKEY"– A SLEIGH-MOUNTED MACHINE WITH DRUMS FOR SPOOLING CABLE– POWERED BY STEAM, GAS, AND LATER, BY DIESEL –

EVERYTHING WENT SMOOTHLY— AL GOT THE PASSLINE CLEARED UP AND TIED OFF TO ONE SIDE, CLEAR OF THE CHAFING OF THE YARDING LINES —

MEANWHILE, ART AND THE REST OF THE CREW ARE PULLING OUT THE STRAWLINE

*"CHASER"— HE UNHOOKS THE CHOKERS FROM LOGS AT SPAR TREE— GENERALLY REQUIRED TO BE A GOOD CABLE SPLICER
*"CUT HER OFF"— UNHOOK--- DISCONNECT
*"THAT'S LINE"— MEANS ENOUGH LINE HAS BEEN PULLED BY HAND —

WHILE THE RIGGING CREW IS STRINGING OUT THE YARDING ROADS, HOWIE, THE *"WHISTLE PUNK" IS TAKING OUT HIS WHISTLE WIRE — HE OPERATES A JERKWIRE WHISTLE, WHICH CONSISTS, SIMPLY, OF A TAUT WIRE STRETCHED FROM THE VALVE ON THE WHISTLE TO A SPRINGPOLE ON THE ROOF OF THE DONKEY, THEN OUT TO THE WOODS — A SPRING HOLDS THE VALVE CLOSED AND ANY JERK ON THE TAUT WIRE OPENS THE VALVE AND OPERATES THE WHISTLE —

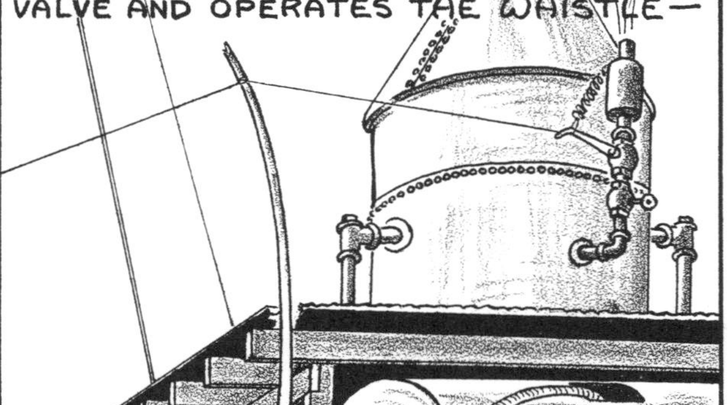

BEING A RATHER INGENIOUS LAD, HOWIE HAS INCORPORATED SOME OF HIS OWN IDEAS INTO THE SYSTEM — HE HAS THE WIRE ON A SPOOL WHICH IS MOUNTED ON THE FRAME OF A PACKBOARD — AS HE WALKS OUT FROM THE DONKEY THE WIRE UNWINDS, CONTROLLED BY HIS GLOVED HAND —

HE HAS THREADED A NUMBER OF OLD, PORCELAIN INSULATORS ON THE WIRE, AND EACH HAS A PIECE OF CORD ATTACHED — WHEN A LENGTH OF WIRE HAS UNWOUND FROM THE SPOOL, HOWIE TAKES OFF HIS PACK, AND GOES BACK AND STRINGS UP HIS WIRE, ATTACHING THE INSULATORS TO ANY CONVENIENT SAPLING, ETC. —

REACHING A VANTAGE POINT, FROM WHICH HE CAN BOTH SEE AND HEAR THE RIGGING CREW, HOWIE TAKES THE PACK FROM HIS BACK, WINDS THE WIRE TIGHT WITH THE CRANK ON THE SPOOL AND JAMS THE CONTRAPTION BETWEEN ANY CONVENIENT OBJECTS TO KEEP THE LINE TIGHT —

WHEN HOWIE WANTS TO "PICK UP" HIS WHISTLE WIRE HE JUST REVERSES THE PACK, SO THAT THE SPOOL IS IN FRONT OF HIM AND WINDS THE LINE UP, UNTIEING THE INSULATORS AS HE COMES TO THEM —

AND NOW, HOWIE HAS HIS WHISTLE WIRE STRETCHED OUT, THE CHASER HAS COUPLED THE HAULBACK TO THE BUTT RIGGING ON THE END OF THE MAINLINE, AND, TRUE TO ART'S EARLIER PREDICTION, THE CREW IS LOGGING IN HALF AN HOUR —

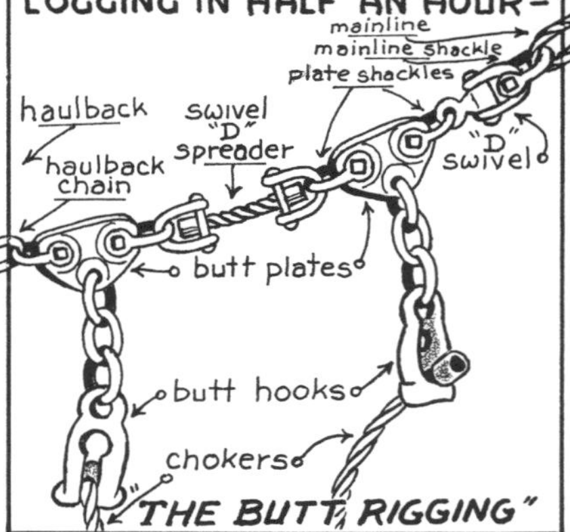

"THE BUTT RIGGING"

LOGGING TERMS

*"WHISTLE PUNK" SIGNAL MAN ON YARDING CREW — HE RELAYS THE SHOUTED ("SCREAMED") SIGNALS FROM THE HOOKTENDER AND RIGGING SLINGER, IN TO THE ENGINEER ON THE DONKEY, BY MEANS OF HIS WHISTLE WIRE —

Bus Griffiths —

THREE JETS OF STEAM, CLOSELY FOLLOWED BY TWO MORE, SHOOT SKYWARD AS THE SHRILL WHISTLES PIERCE THE AFTERNOON AIR, AND ECHO ALONG THE HILLS— THERE IS A SHARP BURST OF POWER FROM THE BIG YARDER AS "SLACKLINE," THE ENGINEER, OPENS THE THROTTLE, AND THE LINES SPRING CLEAR OF THE BRUSH, TO STAND QUIVVERING FOR A MOMENT, LIKE STEEL RIBBONS ETCHED AGAINST THE SKY--- THEN THE LINES ARE SLACKED DOWN AND DAN, THE CHASER, CLICKS THE CHOKER KNOBS INTO THE BUTT HOOKS—

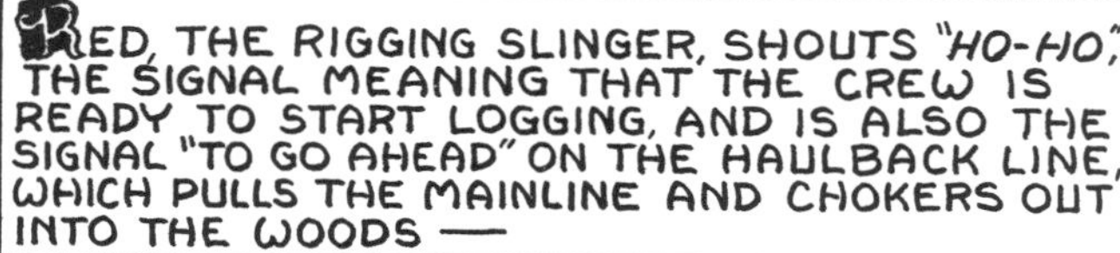

RED, THE RIGGING SLINGER, SHOUTS "HO-HO", THE SIGNAL MEANING THAT THE CREW IS READY TO START LOGGING, AND IS ALSO THE SIGNAL "TO GO AHEAD" ON THE HAULBACK LINE, WHICH PULLS THE MAINLINE AND CHOKERS OUT INTO THE WOODS —

LOGGING TERMS *"TIGHTLINE"— SIGNAL FOR ENGINEER TO "GO AHEAD" ON BOTH MAIN AND HAULBACK— THIS TIGHTENS ONE LINE AGAINST THE OTHER, AND "STANDS THE RIGGING UP," CLEAR OF ALL OBSTRUCTIONS

THE CHOKERS SAIL OUT, AND RED, THE RIGGING SLINGER, BARKS HIS SIGNALS TO THE WHISTLE PUNK—WHITE JETS OF STEAM SHOOT SKYWARD, TO HANG LIKE MISTY PLUMES AGAINST THE DARK TRUNK OF THE SPAR TREE, AND THE SHRILL WHISTLES ECHO ALONG THE HEAVILY TIMBERED HILLS— THE CLINK OF METAL ON METAL CAN BE HEARD AS THE CHOKERMEN DRAG OUT THE STEEL CHOKERS AND SET THEM ON THE BIG BROWN LOGS—

A SINGLE SHOUT FROM RED, ONE SHRILL BLAST FROM THE WHISTLE, AND A JET OF STEAM SHOOTS UP— THERE IS A BURST OF POWER FROM THE BIG YARDER AS 'SLACKLINE' OPENS THE THROTTLE— TWO LOGS LEAP FROM THE BRUSH, CRASHING DOWN SNAGS, UPROOTING SAPLINGS, BOUNCING OFF STUMPS, AS THEY FOLLOW THE PULL OF THE MAINLINE AND BUMP AND SLIDE IN TOWARDS THE SPAR TREE- THE TURN REACHES THE LANDING AND THE SHARP JANGLE OF THE BUTT RIGGING CAN BE HEARD AS THE ENGINEER SLACKS THE LINES- THERE IS A SLIGHT PAUSE, AND THE CLINK OF METAL, AS THE CHASER GOES OUT TO UNHOOK THE TURN, THEN THE QUIETER SOUND OF THE HAULBACK WORKING AS THE CHOKERS SAIL OUT TO THE WOODS AGAIN—

ALL AFTERNOON THE CHOKERS SAIL OUT AND THE LOGS THUMP AND CRASH IN TO THE LANDING, UNTIL, FINALLY----
TOOT-TOOT-TOOT--TOOT-TOOT-TOOT--TOOT-TOOT--TOOT
THERE'S THE *"SLACKOFF! LET'S GO, BOYS!

YOU RIDE WITH ME IN THE PICKUP, AL— SURE HOPE WE DON'T GET HELD UP TOO LONG WITH 'EM LOADIN' OUT AT THE OLD SETTIN'
MAYBE WE'LL BE LUCKY, ART! THEY MIGHT BE ALL FINISHED!

WE'RE IN LUCK! "CHILLIWACK" IS PUTTIN' THE *"PEAKER" ON BARNEY NOW- GOD! THAT'S ONE HELL OF A LOAD!!!

BARNEY WHEELS THE BIG TRUCK OUT OF THE LANDING, AND STOPS OPPOSITE THE *"TURNAROUND" TO ALLOW THE FOREMAN'S PICKUP AND THE CRUMMY, BEARING THE REST OF THE YARDING CREW, TO PASS-

THAT KNOTHEAD MUST HAVE DRY ROT BETWEEN HIS EARS, THROWIN' A LOAD LIKE THAT ON OL' BARNEY---I THINK I'LL JUST STOP A MINUTE AND SMARTEN THAT BASTARD UP!

LOGGING TERMS

*"*SLACKOFF*"— QUITTING WHISTLE

*"*PEAKER*"— TOP LOG ON THE LOAD

*"*TURNAROUND*" (*TURNOUT*)— A WIDE SPOT IN THE ROAD WHERE THE TRUCKS CAN TURN AROUND, OR PASS.

BUS Griffiths-

LOGGING TERMS

*"LANDING"— AREA AROUND SPAR TREE WHERE LOGS ARE LANDED AND LOADED OUT

*"HEAD LOADER" HEAD MAN ON THE LOADING CREW— IN A TRUCK SHOW HE STANDS ON CAB OF TRUCK AND PICKS THE LOGS FOR HIS LOAD

*"OPEN SLASH" LOGGED-OFF COUNTRY

Bus Griffiths—

A FEW YEARS AGO A LOT OF THE GUYS SAID TRUCKS WERE NO GOOD IN THE WOODS— THEY FIGURED TRAINS WAS THE ONLY THING—

THE BIG OUTFITS ARE USING *"LOCIES" AND THEY'RE LOGGIN' IN THE BIG VALLEYS—THEY LAY THEIR STEEL IN THE BOTTOM AN' REACH UP THE SIDEHILLS AS FAR AS THEY CAN WITH THE *"SKIDDERS" OR SOME KIND OF BIG STEAM POT USING A SKYLINE SYSTEM— THE TIMBER THEY CAN'T REACH, THEY LEAVE!

MOST OF 'EM LOAD WITH A *"DUPLEX" AN' THEY'RE USED TO SLAMMIN' THE LOGS AROUND- IF THEY KNOCK A CAR OFF THE TRACK THEY JUST GRAB IT WITH A LOADIN' TONG AN' SET IT BACK ON THE TRACK— YOU *SURE AS HELL* CAN'T HANDLE *TRUCKS LIKE THAT!*
NO- THEY'D FALL APART

A LOT OF OUTFITS TRIED TRUCKS, BUT THEY WASN'T GETTIN' ANYWHERE— THEY THOUGHT THEY COST TOO MUCH, BROKE TOO EASY, AN' WOULDN'T PACK ENOUGH WOOD—BUT THEY WAS ALL FOOLIN' AROUND WITH SINGLE-AXLE JOBS— THEN SOME OLD FARMER FROM THE FRASER VALLEY STARTED HAULING LOGS OFF VEDDER MOUNTAIN WITH TRUCKS, AN' HE SHOWED EVERY-BODY HOW TO USE 'EM—

I GUESS HE WAS A MECHANIC, AN' A GOOD ONE— ANYWAY, HE MADE SIX-WHEELERS OUT OF HIS TRUCKS AN' THAT WAS THE ANSWER!

THE BACK AXLE DIDN'T DRIVE, BUT IT MADE THEM INTO A BETTER TRUCK, AN' THEY'D HAUL NEARLY TWICE THE LOAD—THEY'VE COME A LONG WAY SINCE THEN, AN' IT'S JUST A MATTER OF TIME UNTIL TRUCKS TAKE OVER FROM THE RAIL-ROAD SHOWS!
SINCE THE START OF THE DEPRESSION THERE'S LOTS OF *"GYPPOS"- MOST OF 'EM ARE USING TRUCKS!

MEANWHILE, THE BIG TRUCK, WITH ITS HEAVY LOAD OF LOGS, BREAKS OVER THE BROW OF THE LONG HILL ABOVE CAMP, LESS THAN A MILE BEHIND THE CRUMMY, AND THE PICKUP TRUCK, BEARING AL RICHARDS AND ART DONNEGAN—
MY GOD! THE BRAKES WON'T HOLD HER-- SHE'S *TOO HEAVY!*
Bus Griffiths-

LOGGING TERMS

*"*LOCIE*"— A LOGGING LOCOMOTIVE

*"*SKIDDER*"— A YARDING DONKEY WITH A SKYLINE SYSTEM FOR ROUGH GROUND—ABLE TO REACH OUT 1200 ft. ON SQUARE LEAD AND 1600 ft. ON CORNERS OF THE SETTING—

*"*DUPLEX*"— A LOADING DONKEY WITH DOUBLE, REVERSING ENGINES-

*"*GYPPOS*"— REFERS TO SMALL OUTFITS-- SOME WERE GOOD CAMPS, AND SOME WERE WHAT THE NAME IMPLIES--- HAYWIRE FROM THE LOG DUMP TO THE TAIL BLOCKS—

THE PICKUP TRUCK ROUNDS THE CORNER INTO CAMP—
WHY IS THE *"CAT" CLEARING OFF THAT CHUNK OF GROUND, ART ?
THE "OLD MAN" WANTS SPACE FOR MORE BUILDINGS---- LOOKS LIKE DOUG IS GONNA WORK OVERTIME—

ART PULLS IN AT THE SIDE OF THE TRUCK SHED, AND AS AL CLIMBS OUT OF THE CAB, JIM BRADLEY APPROACHES—
HI, JIM!
'LO, AL---SAY, ART, YOU GOT A MINUTE?
SURE, JIM--- CLIMB IN—

MEANWHILE----- WHAT OF BARNEY?
JUST LISTEN TO THAT MOTOR HOWL! SOUNDS LIKE IT'S GONNA FLY OUT AN' HIT ME IN THE FACE!
THE ROAD IS PRETTY FLAT THRU CAMP---IF I CAN JUST HOLD HER ON THE ROAD!
BARNEY CAN FEEL THE RUNAWAY TRUCK LOSING SPEED AS HE NEARS THE CORNER HEADING INTO CAMP—
I BETTER LAY ON THE HORN-- IF I CAN HOLD HER ON THIS CURVE I MIGHT GET HER STOPPED ON THE FLAT STRETCH THRU CAMP!!
*"CAT"—CATERPILLAR TRACTOR—

STRONG ARMS GUIDE THE RUNAWAY TRUCK ON THE CURVE, AND HORN BLARING, THE BIG TRUCK ROUNDS THE CORNER INTO CAMP—
GOOD GOD!! THE CRUMMIES! !?!

UNAWARE OF THEIR DANGER, AND UNABLE TO HEAR THE BLARING HORN ABOVE THE ROAR OF THE *"BULLDOZER" WORKING IN CAMP, THE BANTERING CREWS PILE OFF THE CRUMMIES, WHICH SIT IN THE MIDDLE OF THE ROAD, SQUARELY IN THE PATH OF THE RUNAWAY TRUCK—
BUS Griffiths
*"BULLDOZER"—A TRACTOR WITH AN ANGLE BLADE USED FOR ROAD BUILDING—

CURSING SAVAGELY, BARNEY WHEELS THE LOGGING TRUCK OFF THE ROAD— THE TRUCK CAREENS WILDLY, AND THE "PEAKER," OR TOP LOG, FALLS OFF-

THE BIG LOG HITS THE GROUND BEHIND THE PICKUP, AND ONE END SWINGS UP HITTING THE BACK OF THE TRUCK—

THE SMASHING BLOW DRIVES THE PICKUP TRUCK THRU THE WALL OF THE TRUCK SHED-

THERE IS A MUFFLED EXPLOSION, AND THE BUILDING IS ON FIRE —

THE LOGGING CREW STANDS IMMOBILE, STUNNED BY THE SUDDEN, DRAMATIC EVENTS OF THE PAST FEW MOMENTS — AL, WHO HAD JUST LEFT THE PICKUP TRUCK, AND HAD STOPPED NEAR THE CRUMMIES TO TALK TO HIS PAL, RED, IS THE FIRST TO MOVE —
RED-- FOR GOD'S SAKE, C'MON! JIM AND ART ARE IN THE FIRE!

THEN THE CREW EXPLODES INTO ACTION
I'LL GET THE FIRST-AID MAN!
WE BEST MOVE THEM BLOODY CRUMMIES!
WE BETTER CHECK ON OL' BARNEY!
C'MON-LET'S GIT THEM FIRE HOSES!
Bus Griffiths

*"GREASE PIT"— A PIT DUG INTO THE FLOOR, MADE SO A TRUCK CAN STRADDLE IT, AND DEEP ENOUGH FOR A MAN TO STAND UP AND WORK ON THE TRUCK'S UNDERCARRIAGE—

MEANWHILE, THE TIMEKEEPER, WHO ALSO DOUBLES AS FIRST-AID MAN, IS GETTING THE CREW ORGANIZED, AND ISSUING ORDERS LIKE A FIELD MARSHAL—

THE MEN WERE UNCEREMONIOUSLY ROLLED IN THE BLANKETS TO SMOTHER THE FLAMES, BUT MOST OF THE FIRE PROVED TO BE IN THE COATS THEY WERE HOLDING OVER THEIR HEADS FOR PROTECTION—

LOGGING TERMS—
*"WOOD LOG"—A LOG THAT HAS BEEN DEPOSITED IN CAMP, TO BE CUT UP AS FUEL FOR THE COOKHOUSE AND BUNKHOUSES—

THEY'RE ASKING TO SEE YOU BOYS- THAT WAS A *GREAT JOB* YOU DID---- GOING *INTO THAT FIRE* AND *PULLING THEM OUT*---A *TERRIFIC JOB*!!
THANKS! HOW BAD *ARE* THEY, ARCHIE?

I DON'T THINK THEY ARE TOO BADLY BURNED, BUT I'M A LITTLE WORRIED ABOUT THEIR EYES---IN THE FLASH OF THE EXPLOSION THEIR FACES AND EYELIDS WERE SCORCHED— THEY'VE LOST ALL THEIR HAIR, EYE-BROWS, AND LASHES!
I SURE HOPE THE POOR BASTARDS ARE GONNA BE ALRIGHT!

I'VE SOAKED THEIR CLOTHING WITH THAT WARM SALT SOLUTION IN CASE OF BODY BURNS, AND WRAPPED THEM IN BLANKETS TO KEEP THEM WARM— THAT WILL COMFORT THEM UNTIL WE GET THEM TO THE HOSPITAL—BUT YOU HAVE *NO IDEA* WHAT A *DIFFICULT MAN* ART DONNEGAN IS!
WHO SAYS *WE DON'T*!

WHAT ABOUT OL' BARNEY? WAS HE BADLY HURT?
I CAN'T REALLY SAY! THINGS HAPPENED *SO FAST*, AND I'VE BEEN BUSY WITH THE BOSS AND ART--- I HAVEN'T HEARD A *THING* ABOUT BARNEY!

NOW, DON'T TALK TO THEM *TOO LONG*--WE'VE GOT TO GET THEM TO THE HOSPITAL AS *SOON AS POSSIBLE*! ***THE BOYS ARE HERE, ART!***
YOU GUYS ARE *SURE SOMETHIN'*! I THINK I'M SPEAKIN' FOR JIM, TOO, WHEN I SAY I'LL *NEVER FORGET* WHAT YOU *DID*-- WE'D *NEVER* GOT OUT ON OUR OWN--- *HELL*— IT HAPPENED *SO BLOODY FAST*!
ART'S RIGHT, LADS---BUT WE WERE LUCKY— WHEN THE PICK-UP WENT THRU THE WALL IT HIT A POST-- WE GOT THROWN OUT, JUST AS IT EXPLODED--- WE WERE ON FIRE, BUT MADE IT TO THE PIT---WE ROLLED AROUND--AN' THEN YOU COME-- *JUST THANKIN'* SOMEBODY *FOR SAVIN' YOUR LIFE DON'T SEEM TO BE ENOUGH*!
AW, HELL---WE'RE JUST GLAD WE MADE IT IN TIME!
Bus Griffiths—

WHAT TH' HELL HAPPENED, ANYWAY? WE WAS SITTIN' IN THE PICKUP----I STILL HAD THE ENGINE RUNNIN'---AN' THEN I SEEN BARNEY GO FLYIN' BY---THERE WAS A HELLUVA CRASH, AN' WE WAS IN THE SHED, AN' ON FIRE---WHAT TH' HELL HAPPENED TO BARNEY?
GUESS HE HAD A RUN-AWAY---HE COME SNORTIN' AROUND THE CORNER AND THE CRUMMIES WERE UNLOADING THE MEN!

BARNEY JUST WHEELED HER OFF THE ROAD AND A LOG FELL OFF AND HIT YOUR PICKUP----I DON'T KNOW HOW HE MADE OUT----
GOD! HE COULD HAVE WIPED OUT TH' WHOLE CREW! I HOPE THE OLD BASTARD'S OKAY!

HI, AL! I HEARD YOU MENTION BARNEY---HE'S OKAY! I JUST LEFT HIM!
THAT'S SURE GOOD NEWS!
WHO THE HELL IS THAT?

IT'S ME---! CHILLIWACK!
BARNEY RUN THE TRUCK INTO THAT BUSH BACK OF THE SHED, AN' HE NEVER GOT A SCRATCH----BUT THE TRUCK'S BUGGERED UP A LITTLE----
TO HELL WITH THE TRUCK! I'M GLAD OL' BARNEY'S OKAY!

WE PARBUCKLED THE LOAD OFF WITH THE CAT, AN' DOUG PULLED THE TRUCK FURTHER FROM THE FIRE---BARNEY'S GONNA CHECK HER TO SEE HOW BAD SHE'S DAMAGED---HE SAID HE COULDN'T HOLD HER ON THE LONG HILL---GUESS THE LOAD WAS TOO HEAVY---SHE RAN AWAY---
BUS Griffiths

ALL THAT, AN' HE JUST LOST ONE LOG OFF THE LOAD! WHEN I LOAD 'EM, THEY'RE LOADED!

IS THAT ALL HE LOST? WELL, HE COULD HAVE LOST HIS LIFE, AN' HE COULD HAVE WIPED OUT THE WHOLE CREW--- YOU SONOFABITCH---YOU BETTER GET YOURSELF A JOB LOADIN' HAY, 'CAUSE YOU SURE AS HELL AIN'T GONNA BE LOADIN' NO MORE LOGS AROUND THIS OUTFIT-- YOU CAN GET YOUR GODDAMNED TIME IN THE MORNIN'--- SEE ABOUT IT, ARCHIE!
I CERTAINLY WILL, ART!
EASY, ART!

I WISH I COULD SEE WHAT THE HELL'S GOIN' ON! IS THE WHOLE SHITEREE BURNIN' UP?
THE TRUCK AND BLACK-SMITH SHOPS ARE A TOTAL LOSS, BUT THE BOYS KEPT THE REST OF THE BUILDINGS WET DOWN AND STOPPED THE FIRE FROM SPREADING!

WHAT ABOUT THE OTHER TWO TRUCKS? WERE THEY IN TH' SHED? AND WHAT ABOUT THAT *"MASTER MANIAC", AN' THE BLACKSMITH?
THE TRUCKS ARE OKAY, ART-- THEY JUST GOT BACK FROM THE *"LOG DUMP"-- CASEY HAD TROUBLE DUMPING HIS LOAD AN' HELD THINGS UP--- THE MECHANIC AND SMITHY ARE OKAY--- THEY HAD JUST GONE TO THE COOKHOUSE FOR A *"MUG UP"--

BUT WE BETTER GET YOU CHAPS DOWN TO THE VILLAGE AND INTO THE HOSPITAL--
I WISH TO HELL YOU'D *SHUT UP*, ARCHIE--- *NOW LISTEN*, AL-- ALL THAT LOADIN' RIGGIN' STILL HAS TO BE STRIPPED OFF THE OLD SPAR AN' HUNG ON THE NEW TREE---
I'LL TAKE CARE OF IT, ART!

I'LL BE LAID UP FOR A FEW DAYS, SO WE'LL SHUT DOWN TH' LOGGIN' 'TILL I GET BACK-- BUT I'LL GET WORD TO CAMP--- LOOK AFTER THINGS WHILE I'M GONE--- DOGGONE IT--- YOU GUYS---- I---
SURE, ART-- YOU GUYS TAKE CARE OF YOURSELVES, AN' GOOD LUCK!
GO EASY ON THOSE NURSES, ART--- AN' WATCH OL' JIM WITH THAT HEAD MATRON

IT'S *TOO BAD* YOU FIRED "CHILLIWACK"-- I GUESS I'M *PARTLY TO BLAME*--- I WAS RIDIN' HIM *PRETTY HARD* ABOUT NOT KEEPIN' UP WITH TH' YARDIN' CREW!
WHATTA YA MEAN, "TOO BAD"? IF I COULDA SEEN TH' SONOFABITCH I'DA BIT TH' FACE RIGHT OFF HIM--- AN' *LISTEN, YOU OLD GOAT-- IN FUTURE*, WHILE *I'M PUSHIN' CAMP*, I DON'T WANT *YOU* INTERFERIN' WITH TH' CREW-- ALL *YOU* GOTTA DO IS *SIGN* THEM BLOODY CHEQUES-- *I'LL* LOOK AFTER TH' LOGGIN'!!
THEY'VE GOT 'EM ABOARD! *LET'S ROLL!*
ART *SOUNDS* OKAY, BUT I DOUBT THAT HE'LL BE BACK IN A FEW DAYS!
YEAH-- IT DON'T SEEM LIKELY-- BUT YOU *NEVER KNOW* WITH GUYS LIKE *HIM*--- THEY SURE DON'T COME *ANY TOUGHER! WE'LL* JUST HAVE TO WAIT AND SEE--- C'MON, LET'S GET SOME SUPPER
LOGGING TERMS
*"MASTER MANIAC"-- A SARCASTIC NAME FOR THE HEAD MECHANIC
*"LOG DUMP"-- WHERE THE LOGS ARE UNLOADED AT THE BOOMING GROUND
*"MUG UP"-- HAVING A CUP OF COFFEE
BUS GRIFFITHS

WELL, THAT'S THAT! WHEN JIM AN' ART GET BACK, WE'RE READY TO GO--YOU MADE GOOD TIME STRIPPIN' THAT OLD SPAR AN' GETTIN' ALL THE LOADIN' RIGGIN' HUNG ON THIS TREE!
THANKS, SLACKLINE-- THINGS WENT PRETTY GOOD! I'LL TURN THE CREW'S TIME IN TO ARCHIE AN' SEE IF THERE'S ANY NEWS ABOUT JIM AND ART—

LATER—AT CAMP —
I GOT WORD THAT JIM AND ART WERE TAKEN TO VANCOUVER FOR FURTHER TREATMENT----THEY WON'T BE BACK FOR AWHILE, SO WE'LL BE SHUTTING DOWN---ANY OF YOU BOYS THAT WANT TO GO TO TOWN CAN GET YOUR CHEQUES!
THANKS, ARCHIE- I'LL TELL THE GUYS!

WELL, RED- IT LOOKS LIKE MOST OF THE BOYS ARE HEADING TO TOWN! HOW ABOUT YOU?
DOGGONE IT, AL - THE WEATHER'S TOO NICE! I SURE HATE BEING IN THE BIG CITY WHEN THE WEATHER IS GOOD!

THAT'S THE WAY I FEEL, TOO! HOW WOULD YOU LIKE TO GO EXPLORING UP THE SOUND? I ALREADY SPOKE TO LEITH ANDERSEN AN' HE SAID HE'D LEND US HIS INBOARD BOAT!
SWELL- AN' WE CAN TAKE OUR BLANKETS, AN' GET SOME GRUB AN' STUFF FROM THE COOKHOUSE!

WE'LL GET EVERYTHING ROUNDED UP, AND WE CAN GET AWAY EARLY IN THE MORNIN'!
NOT TOO EARLY, I HOPE!

AND SO---THE NEXT MORNING—
IT'S DAYLIGHT IN THE SWAMP, RED- RISE AND SHINE!
YOU GOT NO HEART- YOU'RE A BORN SLAVE DRIVER!
OFFICE
Bus Griffiths

Dawn finds our two friends heading out from the booming ground, the wake from the little inboard boat making silver ripples on the dark, quiet waters of the sound—

LEITH LEFT A HANDLINE AN' A SALMON SPOON IN THE BOAT--- YOU KNOW WHAT THE OLD FISHERMEN SAY, RED- 'IF YOU POLISH THE SPOON ON YOUR SHIRT TAIL, YOU'RE *SURE* TO CATCH A NICE FEMALE SALMON'!

BOY! THIS IS THE LIFE, RED!
YEAH—IT SURE IS! ONLY *ONE THING WRONG*--- IT'S *TOO DAMNED EARLY!*

WOW! **HOLD 'ER!** KILL THE ENGINE I'VE **GOT ONE!**--- FEELS LIKE A **BIG SPRING!**

AL COULD FEEL THE FISH FIGHTING DEEP DOWN, AT THE END OF THE LINE— SLOWLY, HE WORKED THE FISH CLOSE TO THE SIDE OF THE BOAT, AND COULD SEE THE WIDE HEAD AND HEAVY BODY, AND KNEW HE WAS HOOKED INTO SOMETHING REALLY WORTHWHILE—

AL WATCHED THE CHANGING COLORS IN THE DAWN'S EARLY LIGHT, AS THE BIG FISH MOVED IN THE CLEAR WATER BENEATH THE BOAT--- THE DARK, SPOTTED BACK, THE BRONZE, BLENDING TO PINK, AND THE SUDDEN FLASH OF SILVER, AS, WITH A SWING OF ITS BROAD TAIL, THE BIG SPRING SALMON TURNED AND STARTED ON A LONG, HARD RUN—
Bus Griffiths—

AL NOTED THAT THE FISH WAS WELL HOOKED AS HE SNUBBED THE LINE THRU HIS FINGERS TO CHECK THE FISH'S RUN—IT FOUGHT OUT NEARLY TO THE END OF THE LINE BEFORE HE GOT IT TURNED, AND THEN IT CAME ALMOST EASILY—

HE WORKED THE BIG FISH ALONGSIDE THE BOAT, AND AS THE BROAD HEAD CAME TO THE SURFACE HE SWUNG THE GAFF HOOK—WITH ONE FLOWING MOTION, AL STRUCK THE FISH ON THE HEAD AND SANK THE GAFF HOME—

FULL DAYLIGHT CAME AS THE LITTLE INBOARD BOAT CHUGGED UP THE SOUND—THE SUN ROSE, BLOOD RED, ABOVE THE DARKLY TIMBERED HILLS, AND WITH THE RISING SUN CAME A BANK OF DARK, RED-TINTED CLOUDS—AL FEELS THE FIRST PUFF OF WIND, HERALDING THE COMING STORM

Bus Griffiths

THE STORM HIT LONG BEFORE THE ROCKY POINT WAS REACHED—FIRST THE LITTLE WIND RIFFLES BREAKING UP THE SURFACE OF THE LONG SWELLS, THEN THE WIND FRESHENED AGAINST THE FLOOD TIDE FLOWING UP THE SOUND, AND THE SWELLS BECAME A ROUGH, UNTIDY SEA—

I DON'T LIKE *THIS* ONE LITTLE BIT!
IT *IS* GETTIN' *PRETTY LUMPY*, ALRIGHT!

I'M KINDA ASHAMED TO ADMIT IT, BUT I'M GETTIN' SCARED, AL!
IT'S NOTHIN' TO BE ASHAMED OF, RED— I DON'T LIKE THIS EITHER---THESE GUYS THAT SAY THEY HAVE NEVER BEEN SCARED IN A STORM ARE EITHER TOO DUMB AND STUPID TO GET SCARED, OR ELSE THEY'RE AWFUL LIARS!

THE MAIN THING IS NOT TO GET PANICKY---YOU CAN BE SCARED, BUT AS LONG AS YOU DON'T LOSE YOUR HEAD YOU ARE IN BETTER SHAPE TO COME THRU THAN THESE FOOLHARDY GUYS---BUT YOU'VE *GOT TO HAVE LOTS OF RESPECT FOR THE SEA!*

WE'RE GAINING ON HER, RED! I BELIEVE WE'RE GONNA MAKE IT!
I SURE AS HELL HOPE YOU'RE RIGHT---IT'S TAKIN' A LONG TIME TO REACH THAT POINT!

AFTER WHAT SEEMED HOURS TO AL AND RED, THE LITTLE BOAT GAINS THE ROCKY POINT AND ROUNDS INTO THE EASIER WATERS OF THE BAY—
LOOK! THERE'S A HOMESTEAD UP AT THE SIDE OF THE BAY!
YEAH, AN' SOMEBODY IS COMIN' DOWN TO THE SHORE! GUESS THEY WERE WATCHIN' US BATTLE THE STORM!

THERE MUST HAVE BEEN SOME KIND OF BIG OUTFIT IN HERE!
YEAH—PROBABLY AN OLD SHINGLE BOLT CAMP---IT LOOKS LIKE THE REMAINS OF AN OLD FLUME, OVER THERE!
Bus Griffiths

YOU BOYS HAD A ROUGH GO OUT THERE! WE WERE UP ON THE BLUFF, AND WE'VE BEEN WATCHING YOU FOR OVER AN HOUR!
I--- I WAS SCARED!
YOU'RE NOT THE ONLY ONE! IT WAS GETTIN' DOWNRIGHT DRAFTY OUT THERE!
YOU'RE NOT KIDDING!

ARE YOU LADS FROM AROUND HERE ?
WE WORK AT BRADLEY LOG, FURTHER DOWN THE SOUND— I'M RED HARRIS, AND HE'S AL RICHARDS—
HI!
H-HELLO!

THIS IS MY DAUGHTER, DEBRA! I'M JOE BROWN---YOU LADS COME UP TO THE HOUSE--- YOU'VE HAD A ROUGH MORNING—MA SHOULD HAVE SOME OF HER GOOD COFFEE ON THE STOVE, AND YOU'LL HAVE LUNCH WITH US!
THANKS--WE DON'T WANT TO PUT YOU OUT NONE---BUT IF YOU INSIST, I'D LIKE YOU TO HAVE A FISH!

RED AND I COULDN'T USE A FISH THIS SIZE, ANYWAY---SO YOU BETTER TAKE IT, JOE!
WHERE DID YOU CATCH THAT MONSTER?
NEAR THAT ROCKY ISLAND, ABOUT NINE MILES DOWN THE SOUND-
OOOH, IT'S BIG!

LATER
MISSUS BROWN, RED AND I WANT TO THANK YOU AND YOUR DAUGHTER VERY MUCH FOR THAT NICE LUNCH--- AN' I'VE NEVER TASTED BETTER COFFEE!
WHY, THANK YOU! YOU'RE MORE THAN WELCOME---AND THANK YOU FOR THAT LOVELY BIG SALMON!

SAY, JOE—WHEN WE WERE COMIN' IN WE NOTICED ALL THE OLD PILING, AND THE REMAINS OF AN OLD FLUME, OR SOMETHIN'..
YOU'RE RIGHT! IT WAS A FLUME, AND A DANDY! THERE WAS A BIG SHINGLE-BOLT CAMP IN HERE, AND THAT FLUME WAS OVER FIVE MILES LONG!
Bus Griffiths.

IS THERE MUCH OF THE OLD STUFF LEFT?
WELL, THERE'S NOT MUCH LEFT OF THE FLUME OR BUILDINGS-- WE GET SOME HEAVY SNOWS AT TIMES--- BUT THE SKIDROADS ARE STILL OKAY!
WE'D SURE LIKE TO LOOK AROUND!
I WORKED HERE BEFORE THE WAR AS A *"TEAMSTER," AND THEN AGAIN WHEN I CAME BACK! SHE SHUT DOWN IN '26 AND I WENT TO VANCOUVER-- WHEN THE DEPRESSION HIT, THINGS GOT TOUGH! THIS PLACE WAS ALWAYS GOOD TO ME, SO I CAME BACK AND TOOK OUT A HOMESTEAD--- I'LL SHOW YOU WHAT THERE IS, AND TELL YOU THE REST!
CAN I COME, DADDY?
WHEN WE CAME HERE, THE TIMBER STARTED AT THE BEACH--- CEDARS LIKE YOU WOULDN'T BELIEVE!
THE FIRST THING THEY DID WAS BRING IN CHINESE IMMIGRANT LABORERS AND BUILD A CAMP ON THE BEACH OF HAND-SPLIT CEDAR—THE GRAIN IN THE TREES WAS SO STRAIGHT THEY COULD SPLIT A 24-FOOT LOG WITH WEDGES, AND MAKE THE LUMBER THEY NEEDED!
THEN THEY STARTED TO BUILD THIS SKIDROAD FROM THE BEACH TO THE BACK END OF THE CLAIM, A DISTANCE OF OVER FIVE MILES!
THEY SURE DID A FINE JOB!
THE ROADS WERE CERTAINLY WELL MADE—THE GROUND WAS GRADED BY HAND-- PICK, SHOVEL, AND MATTOCK, THEN THE CUT TIMBER WAS LAID-THE STRINGERS WERE SET IN PLACE, AND EVERY FOUR FEET A PEELED SKID WAS DRIFTED TO THE STRINGERS—BETWEEN THE SKIDS, HAND-SPLIT *"CEDAR PUNCHEON" WAS LAID—
LOGGING TERMS
*"TEAMSTER"— A MAN WHO DRIVES HORSES-
*"CEDAR PUNCHEON" ROUGH-SPLIT PLANKS, WHICH MADE A DRY, AND EVEN SURFACE FOR THE HORSES TO WALK UPON-GENERALLY SPLIT FACE GRAIN, AS LESS LIKELY TO CRACK THAN EDGE GRAIN PLANKS—
BUS Griffiths

THE ODD CLUSTER OF SHAKE SHACKS WAS BUILT AS THE ROAD PROGRESSED UP THE VALLEY, BUT NO PERMANENT CAMPS— THAT CAME LATER —

ONCE THE MAIN SKIDROAD WAS FINISHED, A PORTABLE MILL WAS BROUGHT IN— IT WAS LOADED ON A BIG SLEIGH, AND HAULED TO THE BACK END OF THE *"CLAIM"—

A SITE WAS PICKED ON THE RIVER AND THE MILL WAS SET UP— THEY BUILT A WING DAM AND MADE A POND— THERE WERE LOTS OF BALSAM AND HEMLOCK TREES GROWING IN-BETWEEN THE CEDARS, AND THESE WERE FELLED, BUCKED, AND HAULED TO THE MILL POND—

AND THEN THEY STARTED CUTTING LUMBER— THE BUILDING OF THE FLUME WAS STARTED FROM THE END OF THE MILL— ONCE A FEW SECTIONS WERE BUILT, THEY TURNED THE WATER INTO IT, AND FLOATED DOWN THE LUMBER AS IT WAS NEEDED —

THE FLUME WAS WELL BUILT, AS YOU CAN SEE! MOST OF THE SUPER-STRUCTURE WAS CUT FROM THE WOODS, BUT THE FLUME ITSELF WAS MADE OF SAWN LUMBER— THERE WAS A WALKWAY RUNNING ITS FULL LENGTH SO IT COULD BE EASILY PATROLLED!
IT WAS QUITE A FLUME, ALRIGHT!
BUS Griffiths

WHEN THE FLUME REACHED THE SITE PICKED FOR THE FIRST CAMP, THE LUMBER WAS CUT IN THE MILL, FLUMED DOWN, AND THE CAMP BUILT

LOGGING TERMS } *"CLAIM"— *PATCH OF SURVEYED TIMBER TO BE LOGGED OFF* —

CHINESE BOLT CUTTERS WERE BROUGHT IN AND THE CUTTING STARTED— WHEN FALLING A TREE, THEY FIRST MADE A SAW CUT ON THE UNDERCUT-SIDE, AND THE SNIPE WAS CHOPPED UP FROM THE UNDERSIDE TO MEET THE SAW CUT— THIS WAS DONE SO THERE WOULD BE A SQUARE CUT ON THE BUTT OF THE TREE, AND THEREFORE NO WASTE IN THE BUTT BLOCK—
THE BLOCKS WERE CUT 56 INCHES LONG AND SPLIT WITH HAMMER, STEEL WEDGES, AND A SPLITTING BAR— EACH BOLT WAS ABOUT FOUR FEET IN CIRCUMFERENCE— THE BOLTS WERE STACKED ON END SO THEY COULD BE TALLIED— IT TOOK ABOUT 32 BOLTS FOR A CORD, AND THE CUTTER RECEIVED 75 CENTS PER CORD—
WHILE THE CUTTING WAS GOING ON, MORE SKIDROADS WERE BEING BUILT THRU THE TIMBER— THE BUILDING OF THE FLUME WENT AHEAD, AND TWO MORE CAMPS WERE BUILT— EACH CAMP HAD BARNS WITH STALLS FOR 30 TO 40 *"HAY BURNERS"
FINALLY THE FLUME WAS FINISHED AND THE PILING DRIVEN FOR THE BOOMING GROUNDS— 16-FOOT SLEIGHS WERE BUILT, WITH RUNNERS MADE OF GUMWOOD, AND EACH SLEIGH COULD HAUL TWO CORDS OF BOLTS— NOW THEY WERE READY TO START HAULING OUT THE SHINGLE BOLTS!
THE SKIDROADS WERE BUILT SO THAT TWO OR THREE ROADS WOULD CONVERGE AT ONE SPOT ALONG THE FLUME— THESE WERE LANDINGS, AND PLACED WHEREVER THE TOPOGRAPHY OF THE LAND WAS CLOSE TO THE LEVEL OF THE FLUME— THE LOADED SLEIGHS WERE DRIVEN ALONGSIDE, AND PARALLEL TO THE FLUME—
MEN WITH PICAROONS PULLED THE SHINGLE BOLTS OFF THE LOADED SLEIGHS INTO THE FLUME, AND THEY WERE ON THEIR WAY TO THE *"SALT CHUCK"— OVER 1000 CORDS OF BOLTS HIT THE CHUCK EVERY WEEK—
BUS GRIFFITHS
LOGGING TERMS
*"SALT CHUCK"— SALT WATER-TIDEWATER
*"HAY BURNERS"— HORSES

THE SHINGLE BOLTS WERE LOADED ABOARD CRIBS—A JACK-LADDER, POWERED BY AN OLD AIR-COOLED ENGINE, WAS USED TO LIFT THE BOLTS OUT OF THE WATER—
A CREW ABOARD THE CRIB STACKED THE BOLTS, AND IT WAS A VERY EXACTING JOB—GREAT CARE HAD TO BE TAKEN TO LOAD THE CRIB EVENLY, OTHERWISE IT WOULD FLIP OVER, AND DUMP THE LOAD— THE MEN WOULD GO FOR AN INVOLUNTARY SWIM, AND THERE WAS ALWAYS THE DANGER OF BEING TRAPPED AND CRUSHED IN THE MASS OF BOLTS
THE LOADED CRIBS WERE TOWED AWAY BY A BIG STEAM TUG, BOUND FOR THE MILLS SCATTERED ALONG THE BANKS OF THE MIGHTY FRASER RIVER—HERE THE BOLTS WERE CUT INTO SHINGLES, PACKED INTO BUNDLES, AND SOLD TO COVER THE ROOFS OF HOUSES THROUGHOUT THE WORLD—
THESE ARE THE BUILDINGS OF CAMP ONE---THEY HAVE LASTED BECAUSE THE SNOWS AREN'T QUITE AS HEAVY HERE AS THEY ARE FURTHER UP THE VALLEY!
THIS WAS *SOME OUTFIT* WHEN IT WAS GOING FULL OUT---OVER 200 CHINESE WORKING, BESIDES THE WHITE TEAMSTERS AND BOSSES-- AND ABOUT 100 HORSES!
MUST HAVE BEEN A BUSY PLACE!
THANKS FOR SHOWIN' US ALL THIS, AN' TELLIN' US WHAT IT WAS LIKE--- IT SURE WAS INTERESTIN'!
THE WAY YOU TOLD IT, MADE IT SEEM LIKE WE WAS BACK THERE WITH THE CHINAMEN!
YES, IT WAS QUITE A PLACE! I CAN'T WALK TOO FAR ANYMORE, BUT I LIKE TO WALK THE OLD SKID-ROADS---IT BRINGS BACK LOTS OF MEMORIES!
Bus Griffiths

THERE'S LOTS OF THESE OLD GHOST CAMPS ALONG THE COAST, AN' THEY'RE ALL FULL OF MEMORIES--- FOR SOMEBODY—IT'S TOO BAD SOME OF THEM COULDN'T BE PRESERVED, SOMEHOW!
YES, IT IS A SHAME!

HELLO! WHAT'S THIS ?

WHAT IS IT, ANYWAY?
OH, THAT'S A CHINAMAN'S SANDAL!

THEY MADE THEM OUT OF CEDAR—THEY SPLIT A PIECE FOR THE SOLE, FASTENED ON A COUPLE OF HEAVIER PIECES FOR CLEATS, AND THEN TACKED A BAND OF LEATHER ACROSS THE TOP TO HOLD THE THINGS ON THEIR FEET! THEY USED TO CLICK-CLACK AROUND CAMP IN THEM!

THEY CARVED PIPES OUT OF BROAD-LEAF MAPLE, AND THEY WERE SURE SWEET-SMOKING PIPES—I STILL HAVE THE ONE THAT A FELLOW CARVED FOR ME OUT OF A HUCKLEBERRY ROOT! THEY WERE INGENIOUS LITTLE PEOPLE AND MADE A LOT OF THE THINGS THEY NEEDED!

THE POOR BEGGARS WERE NEVER PAID MUCH MONEY---- LITTLE MORE THAN THEIR GRUB!
I GUESS THAT'S WHERE THE SAYIN' CAME FROM ABOUT "WORKIN' FOR CHINAMENS' WAGES"!
I'D LIKE TO FIND ANOTHER ONE OF THESE THINGS--- IT WOULD BE NICE TO HAVE A PAIR, JUST TO KEEP!

W-WE-- WE HAVE QUITE A FEW AROUND HOME---I-I'LL GIVE YOU A PAIR--- IF YOU'D LIKE!
WHY, THANK YOU! I'D LIKE THAT!
BUS Griffiths

WE'D BEST BE GETTING BACK! FROM THE LOOK OF THE SKY IT'S STILL ROUGH ON THE WATER— YOU BOYS BETTER STAY— HOPE YOU WON'T MIND SLEEPING IN THE BARN!
NOT AT ALL--THAT WILL BE SWELL!
I DID A LOT OF THAT WHEN I WAS RIDING THE FREIGHTS!

WHAT'S A GIRL LIKE YOU DOIN' IN A PLACE LIKE THIS? DON'T YOU GET LONELY?
W-WHERE ELSE *WOULD* I BE? *THIS* IS *MY HOME!* I DO GET LONELY, AT TIMES, BUT I LOVE IT HERE! IT'S SUCH A PEACEFUL PLACE TO LIVE!

ARE YOU REALLY A LOGGER? YOU--YOU DON'T SEEM LIKE ONE---
I GUESS I'M A LOGGER, ALRIGHT--- BUT WHY DON'T *I SEEM* LIKE ONE?

EVERYONE SAYS LOGGERS ARE AWFUL MEN---TERRIBLY WILD AND ROUGH, AND I SHOULD KEEP AWAY FROM THEM--- BUT YOU'RE NOT LIKE THAT AT ALL---YOU'RE POLITE---AND NICE!
WHY, THANK YOU--I KNOW LOGGERS HAVE A BAD NAME, BUT IT'S NOT FAIR TO LUMP THEM ALL IN ONE BUNCH— MOST OF THE GUYS I WORK WITH ARE FINE MEN---A LOT BETTER THAN MOST!

I KNOW SOME LOGGERS ARE PRETTY CRUDE! MAYBE THE WORK, AND THE ISOLATION IN CAMPS HAS SOMETHING TO DO WITH IT---IT'S A TOUGH, DANGEROUS JOB, AND YOU HAVE TO WORK IN ALL KINDS OF WEATHER--- I GUESS IT BREEDS HARD ROUGH MEN!
W-WHY DO YOU WORK AT IT, THEN?

THAT'S SIMPLE---*I LIKE* LOGGING! I'M NOT ALWAYS PROUD OF WHAT WE'RE DOIN' TO THE COUNTRY, BUT I CAN'T THINK OF ANY KIND OF WORK I'D RATHER DO---TO ME, IT'S MORE A WAY OF LIFE THAN A JOB!
BUS GRIFFITHS

DID YOU NOTICE THAT MOST OF THE BRIDGES ARE BUILT WITH A CURVE IN THEM?--WHY DID THEY BUILD THEM THAT WAY---DO YOU KNOW?
YES, I BELIEVE I DO!

MOST OF THESE BRIDGES WERE BUILT BY HAND, AN' THEY HAD TO USE TIMBERS THEY COULD HANDLE---SO THEY HAD TO MAKE THEM AS STRONG AS POSSIBLE! A STRAIGHT BRIDGE CAN ONLY BE BRACED FORE AN' AFT AN' CROSSWAYS--BUT A CURVED BRIDGE CAN BE BRACED FROM NEARLY EVERY ANGLE, AN' THAT MAKES IT A LOT STRONGER!
OH---IS THAT RIGHT?

SURE-JUST LOOK OVER THE SIDE HERE AN' I'LL SHOW YOU--BUT DON'T TRUST THIS HAND RAIL TOO MUCH---IT MIGHT BE ROTTEN!

HERE---I'LL HANG ONTO YOU!--
OH-H ?!?

DO YOU SEE HOW ALL THOSE BRACES ARE COMING IN AT DIFFERENT ANGLES AS THE BRIDGE CURVES?---THAT'S WHAT MAKES IT SO STRONG!
Y-YES!?!

I'M SORRY !!
I SHOULDN'T HAVE HELD YOU LIKE THAT! BUT I---I DIDN'T MEAN TO BE FRESH----I--I GUESS WE BETTER GO!
Y-YES---I---I THINK SO!

I HOPE YOU'RE NOT MAD AT ME---I SURE WASN'T TRYIN' TO BE FRESH---I DUNNO---MEBBE I'VE BEEN IN CAMP TOO LONG!
I FORGIVE YOU---I HOPE YOU COME BACK SOMETIME---I--I LIKE TALKING TO YOU---BUT WE'RE NEARLY HOME---I--I'LL SEE YOU AT SUPPER!
BUS GRIFFITHS

FOLLOWING A HEARTY SUPPER OF BAKED SALMON, NEW POTATOES, AND GREEN PEAS, TOPPED BY A BIG WEDGE OF APPLE PIE, AND WASHED DOWN BY SEVERAL CUPS OF MRS. BROWN'S DELICIOUS COFFEE, THE MEN RETIRED TO THE LIVING ROOM FOR AN EVENING OF QUIET TALK

DO YOU JUST HAVE THE ONE DAUGHTER, JOE?
OH, NO! I HAVE TWO SONS, ALSO---ONE OLDER, AND ONE YOUNGER THAN DEBBIE---THEY'RE AWAY ON THE BOAT, FISHING—

IT'S BRUCE'S FIRST YEAR, BUT HARV HAS BEEN AT IT FIVE YEARS—HE DOES QUITE GOOD, AND HE SEEMS TO LIKE THE LIFE!
YEAH—IT'S A GOOD LIFE--I FISHED A COUPLE SEASONS, BUT I LIKE THE LOGGING BETTER!
AFTER THAT SESSION OUT THERE TO-DAY, I KNOW I'D SOONER LOG!

AT TIMES I WISH HARV DIDN'T HAVE THE BOAT—I COULD SURE USE HIS HELP AROUND THE PLACE— THERE'S LOTS OF CHORES THAT JUST DON'T GET DONE! DEBBIE IS A BIG HELP, BUT THERE'S LOTS OF THINGS A GIRL JUST CAN'T DO!
I-I'M NOT VERY GOOD AT CHOPPING WOOD!
THAT'S NOT EXACTLY A GIRL'S JOB, ANYWAY!

I DON'T MEAN TO BE RUDE BY CHANGING THE SUBJECT, BUT DIDN'T YOU BOYS SAY YOU ARE LOGGERS?
YEAH—DUMB OLD LOGGERS, THAT'S US!
YOU SPEAK FOR YOURSELF, PARTNER!

I'VE GOT A PATCH OF TIMBER AT THE BACK OF MY PLACE THAT I'D LIKE YOU TO LOOK AT-----IF YOU CAN SPARE THE TIME—
RIGHT NOW WE'VE GOT NOTHING BUT TIME, JOE!

WE CAN START OUT RIGHT AFTER WE HAVE BREAKFAST, IF THAT'S OKAY WITH YOU!
THAT'LL BE FINE--NOW WE'D BETTER GET OUR BLANKETS FROM THE BOAT AND HEAD FOR THE BARN-- THANKS FOR A FINE SUPPER, AND GOODNIGHT!
THANKS, AND GOOD-NIGHT, FOLKS!
GOODNIGHT!
Bus Griffiths

AL AWOKE AT DAYLIGHT, AND, UNABLE TO SLEEP LONGER, HE DRESSED, AND WENT OUT TO THE WATERING TROUGH FOR A BRISK WASH IN THE COLD SPRING WATER—

I'M NOT VERY GOOD AT CHOPPING WOOD

THIS IS NO JOB FOR A GIRL!

SOME TIME LATER
GOOD MORNING! WHY ARE YOU CHOPPING THE WOOD?
HI! OH--- I WAS JUST WORKIN' UP AN APPETITE--NO-- THAT'S REALLY NOT TRUE! I WAS REMEMBERING WHAT YOU SAID LAST NIGHT!

YOU--- YOU DIDN'T NEED TO CHOP IT!
OH, YES I DID! I DIDN'T WANT TO BE THINKING OF YOU HAVING TO CHOP ALL OF THIS WOOD BY YOURSELF!

... AND I WILL BE THINKING ABOUT YOU--- A LOT!!
OH-H-H-H !?!

AL WATCHED AS THE SHY, PRETTY GIRL HURRIED TOWARDS THE BARN— AT THE DOOR SHE PAUSED AND LOOKED BACK, AND AL SAW THAT SHE WAS SMILING!
Bus Griffiths

C'MON, ROMEO— WE'LL NEVER GET THIS WOOD SPLIT IF YOU'RE GONNA STAND THERE MOONIN' ALL DAY!
GOOD OL' RED! LET'S FLY AT IT!

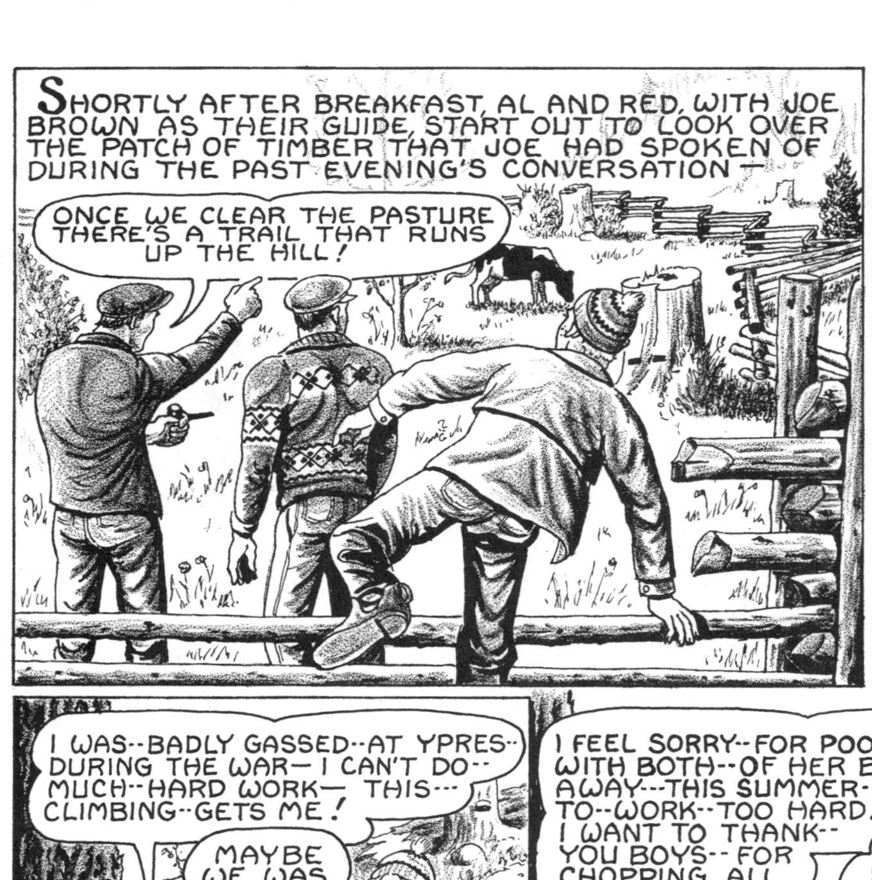
SHORTLY AFTER BREAKFAST, AL AND RED, WITH JOE BROWN AS THEIR GUIDE, START OUT TO LOOK OVER THE PATCH OF TIMBER THAT JOE HAD SPOKEN OF DURING THE PAST EVENING'S CONVERSATION —
ONCE WE CLEAR THE PASTURE THERE'S A TRAIL THAT RUNS UP THE HILL!

WHAT'S WRONG, JOE?
I'LL HAVE---TO REST--AWHILE---LADS---

I WAS--BADLY GASSED--AT YPRES--DURING THE WAR— I CAN'T DO--MUCH--HARD WORK— THIS---CLIMBING--GETS ME!
MAYBE WE WAS WALKING TOO FAST- I'M SORRY, JOE!
I FEEL SORRY--FOR POOR DEBBIE— WITH BOTH--OF HER BROTHERS AWAY---THIS SUMMER--SHE HAS TO--WORK--TOO HARD! I WANT TO THANK--YOU BOYS--FOR CHOPPING ALL THAT WOOD!
DON'T MENTION IT, JOE— WE WERE GLAD TO DO IT!
THE TRAIL WINDS UP THRU AN OLD LOGGING SLASH, AND AN HOUR LATER, THE MEN CROSS A ROCKY RIDGE, AND THE TRAIL PITCHES SHARPLY DOWN-HILL TO ENTER A HEAVILY-TIMBERED DRAW —

WELL---- HOW DOES IT LOOK TO YOU?
YOU'VE GOT A NICE PATCH OF TIMBER, IF IT'S ALL LIKE THIS!
THERE SURE IS SOME FINE TREES HERE!

BUT HOW YOU GOIN' TO GET IT OUT? YOU SURE AS HELL CAN'T DRAG IT UP OVER THIS ROCK BLUFF!
AH! THAT'S THE RUB, LAD- I THOUGHT MAYBE YOU COULD TELL ME!
Bus Griffiths

WELL, JOE—I'D SAY YOU'VE GOT OVER A *"MILLION FEET" OF FIR-- AN' THERE'S A NICE BUNCH OF CEDAR, TOO!
THAT MUCH, EH?--- I DIDN'T---
WOW! LOOK AT THE BIG BUCK!
*REFERS TO A MILLION "BOARD FEET"

BOY! THAT'S A DANDY! HE'S STILL IN VELVET!
THERE'S LOTS OF GAME AROUND— HARV TRAPS IN THE WINTER-- HE HAS A CABIN ON THE OTHER SIDE OF THE DRAW—

THIS IS PRETTY GOOD GROUND HERE! WHAT'S IT LIKE OVER NEAR THE SALT CHUCK?
A STRAIGHT DROP TO THE WATER!

WOW! YOU WEREN'T KIDDIN'--THAT'S A LONG WAYS DOWN! HOW'S THE WATER? DEEP, OR SHALLOW?
SHE'S DEEP-- VERY DEEP!

THAT'S GOOD! BETTER DEEP THAN TOO SHALLOW---I'M BEGINNING TO GET AN IDEA!
I HOPE IT'S A GOOD ONE!

THIS IS GOOD GROUND FOR "CAT LOGGIN'"! WE MIGHT BE ABLE TO MAKE A DEAL WITH OUR BOSS TO GET THE CAT, WHEN WE'RE SNOWED OUT AT THE CAMP THIS WINTER!

WE COULD LOAD THE CAT ON A SCOW, AN' HAVE IT TOWED INTO YOUR BAY, AN' COME INTO THIS DRAW FROM THAT SIDE---WE CAN "FALL AN' BUCK" THIS TIMBER, YARD IT TO THE EDGE AN' PUSH IT OVER WITH THE CAT-- THE LONG DROP WON'T HURT THE LOGS BECAUSE THE WATER IS DEEP!
BUT THAT'S OPEN WATER! THERE'S NO PLACE TO BOOM THE LOGS!
BUS GRIFFITHS
"CAT LOGGING"— TO LOG WITH CRAWLER-TYPE TRACTOR
"FALL & BUCK"—TO CUT DOWN, AND CUT INTO LOG LENGTHS

IT'S OPEN WATER, ALRIGHT— BUT WE COULD JUST PUSH THE LOGS OVER ON CALM DAYS--- DIDN'T YOU SAY YOUR SON HAS A FISHBOAT?
THAT'S RIGHT- A GOOD ONE!
WELL, HE COULD ROUND UP THE LOGS WITH HIS BOAT, AND TOW A STRING OF 'EM INTO YOUR BAY— WE'LL BOOM THEM UP IN THERE!
BY GEORGE, I BELIEVE THAT WOULD WORK! WE SURE COULDN'T FIGURE HOW WE'D EVER GET THIS TIMBER OUT!
NOW, ALL WE'VE GOT TO DO IS FIND A SPOT WHERE WE CAN GET THE CAT DOWN OVER THAT ROCK BLUFF AN' INTO THIS DRAW!
RIGHT, PARTNER! LET'S LOOK AROUND!
YOU GO AHEAD! I'LL REST A SPELL--
THERE DON'T SEEM TO BE A BREAK IN THE ROCK--- IT WOULD BE PRETTY DAMN HAIRY TRYIN' TO GET A CAT DOWN OVER THAT!
IT SURE AS HELL WOULD-- --BUT LET'S KEEP LOOKIN'!
SOME TIME LATER
THAT LOOKS LIKE A BREAK IN THE ROCK FACE, AL!
YEAH--IT MIGHT BE A NARROW DRAW!
WHAT DO YOU THINK, AL?
IT'S PRETTY NARROW, BUT I BELIEVE WE CAN MAKE IT, RED—LET'S GO THRU TO THE TOP!
YOU CAN NEVER GET THE CAT PAST THAT BIG TREE!
Bus Griffiths-

WE'LL HAVE TO FALL THE TREE AND CUT THE STUMP LOW-- THEN THE CAT CAN RUN OVER THE STUMP!
WELL, I'LL BE DAMNED! I GUESS YOU MUST THINK I'M PRETTY STUPID----
NO WAY, JOE! WE'D BE THE STUPID ONES, IF WE THOUGHT THAT!
RED'S RIGHT, JOE! NOBODY IS BORN ON THE JOB---EVERYONE HAS GOT TO LEARN- THIS STUFF HAPPENS TO BE OUR KIND OF WORK!
THE DRAW MAY BE A BIT NARROW IN SPOTS—WE MIGHT HAVE TO BLAST SOME OF THAT ROCK!
IT'LL TAKE A LITTLE WORK, BUT I DON'T THINK WE'LL HAVE ANY TROUBLE GETTIN' THE CAT DOWN THRU HERE!

I'M SURE WE CAN MAKE A DEAL TO GET THE CAT- IF RED IS GAME, AND YOU WANT THE TIMBER LOGGED, I'D LIKE TO COME UP AND GET STARTED WHEN WE'RE SNOWED OUT AT THE CAMP!
WELL, I'M SURE GAME!
THE GOOD LORD MUST HAVE SENT YOU LADS! THINGS HAVE SURE BEEN TIGHT FOR US, BUT IF I CAN SELL SOME LOGS---

LATER THAT DAY AL AND RED ARE REGRETFULLY BIDDING "GOODBYE" TO THE BROWN FAMILY—
YOU FOLKS HAVE BEEN VERY KIND TO US, AND WE SURE HATE TO GO--THANK YOU FOR EVERY-THING---AND WE'LL KEEP IN TOUCH!
YOU LADS DO THAT! AND THANK YOU! MEETING YOU HAS CERTAINLY GIVEN ME NEW HOPE!
GOD BLESS YOU BOTH!

I SHOULD GO DOWN TO THE BOAT WITH YOU, BUT I'M ABOUT DONE IN AFTER THAT WALK!
I'LL WALK WITH THEM, DADDY!
'ER--A--SAY, AL- I'LL GO AHEAD AND TAKE OUR BLANKETS DOWN TO THE BOAT!

SAY, THIS ISN'T THE TRAIL WE USED WHEN WE CAME UP FROM THE BOAT THE OTHER DAY!
N--NO--- THIS IS THE LONG WAY DOWN!
BUS Griffiths-

AS DEBBIE GAVE THE PACKAGE TO AL THEIR HANDS TOUCHED, AND THE CHINESE SANDALS FELL TO THE GROUND, TO LIE AMONG THE FERNS AND MOSSES AT THE EDGE OF THE TRAIL---- FORGOTTEN FOR THE MOMENT----

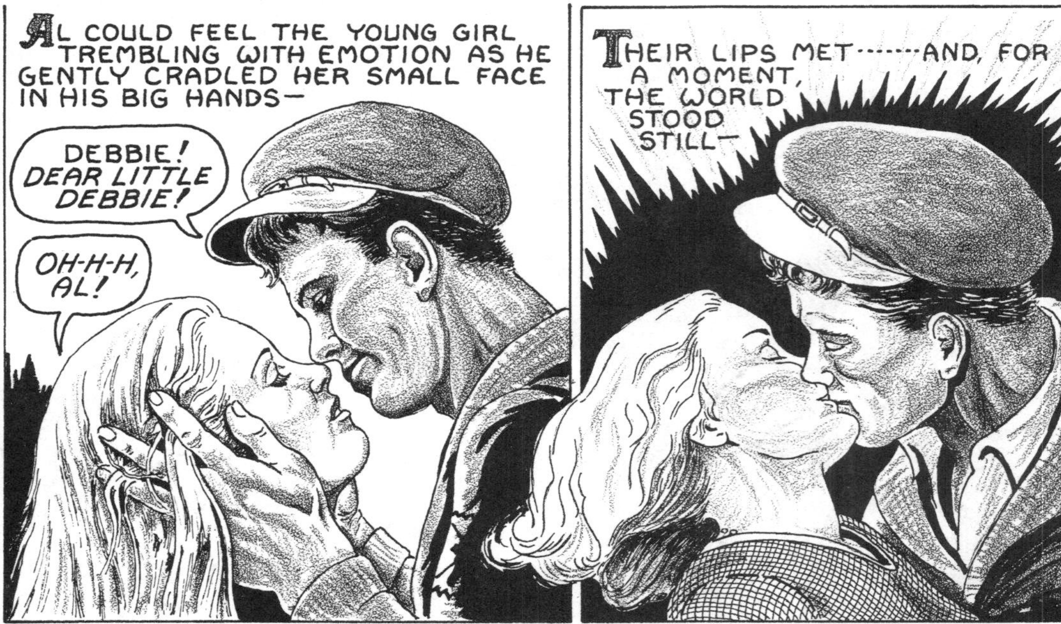

AND, AS THE LITTLE BOAT ROUNDS THE POINT, THE FIGURE OF THE LONELY GIRL FADES FROM SIGHT, AND A GREAT LUMP COMES TO AL'S THROAT--- HE FEELS AS IF HE IS LEAVING A GOOD PART OF HIMSELF BACK AT THE HOMESTEAD WITH THE SHY YOUNG GIRL---AND HE ALSO REALIZES THAT HIS ERSTWHILE CAREFREE, WORK-A-DAY LIFE HAS SUDDENLY BECOME COMPLICATED

AL WAS GLAD TO BE BACK ON THE JOB···THIS WAS HIS WORLD, THE WORLD HE KNEW AND UNDERSTOOD-······HE LEANED CONTENTEDLY AGAINST A STUMP, WATCHING TWO BIG LOGS, AS THEY FOLLOWED THE PULL OF THE MAINLINE, ON THEIR WAY IN TO THE LANDING·············· HE LISTENED TO THE FAMILIAR SOUNDS·····THE GRUNTING AND BUMPING OF THE LOGS AS THEY SLID PAST A STUMP···THE THROATY ROAR OF THE DONKEY, AND THE SLAP AND RATTLE OF GUYLINES AND BLOCKS IN THE STRAINING TREE·····THEN THE SHRILL BLASTS OF THE NOON WHISTLE, AS THE ENGINEER REACHES UP AND JERKS HIS WHISTLE PULL········

WE TURN TO THE TWO MEN UNDER DISCUSSION···JIM BRADLEY, AND HIS CAMP PUSH, ART DONNEGAN·······

WELL, JIM, IT'S GREAT TO BE BACK IN THE *"JUNGLE" AGAIN

YEAH···*SMELL THAT AIR*···IT'S *GREAT* TO BE ALIVE! AN' IT MAKES A GUY REALIZE HOW MUCH WE OWE TO AL AND RED!

LOGGING TERMS

*"JUNGLE"— THE WOODS, AND LOGGING CAMPS IN GENERAL·······

*"COLD-DECK"— LOGS THAT ARE PILED UP, AWAY FROM THE ROAD OR TRACK······· THE COLD-DECK HOOKTENDER IS THE MAN IN CHARGE OF THE MACHINE AND CREW USED FOR THAT PURPOSE

WHAT'S WRONG, PETE ?
PY GOLLY—AYE BANE LOOKIN' FER YOU FELLERS---AYE KVIT DIS YERKVIRE OUTFIT--- DEY OXPECT A MAN T'DO A YOB, AN' PY GOLLY, DEY GIVE HIM A BUNCH OF YUNK TO VORK VID, AN' A BUNCH OF YERKS FER A CREW!

LISSEN, YOU OLD GOAT—I WAS GONNA FIRE YOU OVER A WEEK AGO, BUT WE HAD THAT ACCIDENT--YOU SHOULDA HAD THAT DONKEY ACROSS THAT RIVER---YOU EITHER *DON'T KNOW HOW* TO GET THAT MACHINE ACROSS, OR ELSE YOU'RE *PLAIN SCARED!*
SCARED? YUMPING YIMINEY---AYE'M NOT SCARED! BUT AYE *VUD NOT* TAKE *DAT PILE OF YUNK,* YOU **CALL** A *MACHINE,* ACROSS *DAT RIVER* FER *ALL D'GIRLS* IN SVADEN--AYE'D HAF TO BE CRAY-SEE LIKE A YACKASS!
OKAY—GO GET YOUR CHEQUE---YOU OLD COOT!

WELL, THAT'S THAT! WHAT DO WE DO NOW? WE'VE GOT TO GET THAT DONKEY ACROSS THE RIVER---THERE'S TWO SETTIN'S OVER THERE THAT HAVE TO BE COLD-DECKED
I KNOW *THAT!*

WE'LL BE MOVIN' THE FALLERS IN TO THAT TRACKSIDE SETTIN' ON THIS SIDE OF THE RIVER IN A SHORT TIME, AND IF THOSE TWO SETTIN'S AIN'T PILED UP BY THE TIME THE FALLERS GET THRU, WE'LL BE HELD UP!
FOR GOD'S SAKE---*SHUT UP----I'M THINKIN'!*

YOU KNOW HOW WE WAS WONDERIN' WHAT WE'D DO ABOUT AL, WHEN HENDY COMES BACK? I'M GONNA PUT *HIM* HOOKIN' ON THE COLD-DECKER, AN' HE CAN HAVE A CRACK AT MOVIN' THAT DONKEY ACROSS THE RIVER!
JUST A MINUTE, NOW!

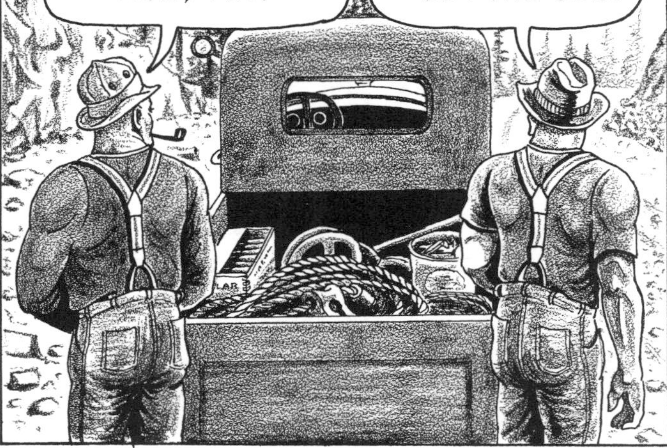
THAT'S A TOUGH JOB FOR AN OLD EXPERIENCED HOOKER, AN' THE LAD'S STILL A LITTLE GREEN-I DON'T THINK HE'S READY FOR ANYTHIN' LIKE THAT, YET!
I AIN'T SO SURE! THE GUY'S KIND OF A NATURAL--- BUT THE ONLY WAY WE'LL FIND OUT IS TO GIVE HIM THE JOB!

WELL, HE'LL LIKELY TACKLE THE JOB, IF YOU ASK HIM--- BUT YOU KNOW WHAT'LL HAPPEN IF HE CAN'T MAKE IT, DON'T YOU?
SURE—*HE'LL QUIT!* BUT, IF HE CAN'T DO THE JOB, HE AIN'T *MUCH USE* TO US---WE'LL HAVE TO TAKE **THAT** CHANCE---'CAUSE I'M ***SURE*** *AS HELL* GONNA ASK HIM---I'LL TALK TO HIM IN THE MORNIN'!
Bus Griffiths

LOGGING TERMS } *"GAS FAKE"—A DONKEY POWERED BY GAS------ THE NAME WAS COINED BY MEN USED TO WORKING WITH STEAM, AND WHO WERE SKEPTICAL OF ANY OTHER KIND OF POWER-----

A SHORT DRIVE THRU A NICE STAND OF TIMBER BRINGS AL AND THE PUSH TO THEIR DESTINATION....

THE COLD-DECK DONKEY SITS AT THE ROADSIDE, NEAR A TALL FIR WITH AN *"X" CHOPPED IN THE BARK— THE CREW IS GATHERED AROUND A STUMP WHERE TWO MEN ARE SPLICING LINE—
JUST WHAT IN HELL HAVE YOU YAHOOS BEEN DOIN' UP HERE FOR THE PAST FEW DAYS? THIS AIN'T NO BLOODY REST HOME, Y'KNOW!

WE TRIED TO KEEP BUSY! WE RUN THE NEW LINES ON THE DONKEY---MADE NEW *"STRAPS" FOR ALL THE BLOCKS---AN' WE'RE JUST FINISHIN' SPLICING THE EYES IN TWO NEW SETS OF GUYLINES FOR THE COLD-DECK TREES!
THAT'S FINE, GEORGE---IT'S MY OWN FAULT.....I SHOULDA COME UP BEFORE AN' SEEN WHAT WAS GOIN' ON!

WHAT WAS MISTER OLSEN DOIN', WHILE YOU BOYS WAS RUNNIN' ON THE NEW LINES, AN' DOIN' ALL THAT SPLICIN'?
HE SPENT LOTS OF TIME DOWN BY THE RIVER-- SAID HE WAS LOOKIN' FOR A CROSSIN' PLACE

I THINK OLSEN WAS SUFFERIN' FROM THE "DUCK'S DISEASE", GEORGE—
OH? WHAT'S THAT?

HE SEEMED TO HAVE TROUBLE GETTIN' HIS ASS UP OFF THE GROUND!
OH---YEAH---I SEE WHAT YOU MEAN....BUT HE DID PICK A PLACE TO CROSS, BELOW THE CANYON.... HE'S GOT SOME TREES MARKED!

BUT HE SAID HE WAS NERVOUS ABOUT SNOOSE RUNNING THE DONKEY, WHEN WE WERE GOIN' ACROSS THE RIVER!

NERVOUS?-- ABOUT SNOOSE WILLIAMS?! LIKE HELL!---THAT GUY WOULD TAKE A DONKEY ANYWHERE! WHY, IF I STOOD ON THE *"HEAD BLOCK" AN' GIVE SNOOSE THE "GO AHEAD" HE'D TAKE HER OVER NIAGARA FALLS! I THINK OLSEN HAS LOST HIS NERVE!

LOGGING TERMS

*"X"— WHEN A TREE HAS BEEN SELECTED AS A SPAR, AN "X" IS CHOPPED IN THE BARK TO PROTECT IT FROM THE FALLERS...

*"HEAD BLOCK"— THE FRONT MAIN CROSS MEMBER OF A DONKEY SLEIGH...

*"STRAPS"— SHORT PIECES OF CABLE WITH AN EYE SPLICED IN EACH END...

Bus Griffiths-

C'MON, AL- LET'S GO SEE THAT CROSSING PLACE OF OLSEN'S!

GOD... SHE LOOKS DEEP!
SHE IS---PLENTY! YOU'LL HAVE TO MAKE DAMNED SURE THAT DONKEY DON'T SINK----BUT THIS IS A POOR BLOODY SPOT TO TAKE HER ACROSS!

WELL, AL- WHAT'S YOUR ANSWER?--- YOU WANT TO TACKLE THE JOB?
TO-MORROW IS SUNDAY, SO WE WON'T BE LOGGIN'--- I'D LIKE A CHANCE TO LOOK IT OVER, AN' THINK ABOUT IT— I'LL GIVE YOU A DEFINITE ANSWER MONDAY MORNIN'!

THAT'S FAIR ENOUGH---BUT I GOTTA KNOW, ONE WAY OR THE OTHER, NO LATER THAN MONDAY!

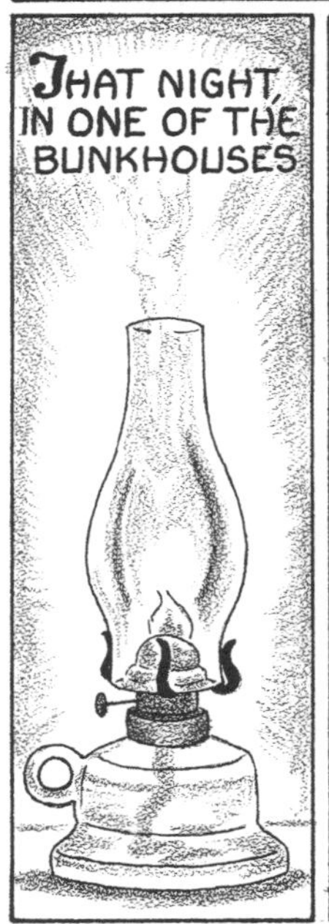
THAT NIGHT, IN ONE OF THE BUNKHOUSES

THAT'S RIGHT, RED---AN' HE WANTED TO KNOW IF I'D TAKE THE HOOKIN' AN' RIGGIN' JOB ON THE COLD-DECKER, AN' MOVE HER ACROSS THE RIVER!
GOD! THAT'S A HAIRY MOVE! ARE YOU GONNA TACKLE IT?
Bus Griffiths

I DON'T KNOW--I'LL LOOK IT OVER IN THE MORNIN'---BUT RIGHT NOW I'M GONNA FORGET ABOUT IT—I'M GONNA SIT DOWN AN' WRITE A LETTER TO DEBBIE--- LIKE I PROMISED
!!!

THE NEXT MORNING
ARE YOU GOIN' UP TO LOOK THINGS OVER AT THE RIVER?
YEAH--THERE'S SURE LOTS OF THINGS I'VE GOT TO TRY AND FIGURE OUT!

SURE YOU DON'T WANT ME TO COME ALONG, AL?
HOPE YOU DON'T MIND, RED, BUT I'D LIKE TO GO ALONE! I WANT TO LOOK THINGS OVER AND MAKE MY OWN DECISIONS-----THEN, IF I MAKE ANY MISTAKES, I'VE JUST GOT MYSELF TO BLAME

ART DIDN'T LIKE OLSEN'S CROSSIN' PLACE...NEITHER DO I... THE FIRST THING IS TO PICK A BETTER SPOT!

LATER- AT THE RIVER...
THERE'S A BEND IN THE RIVER, JUST AHEAD!

WELL, I'M SURE NO EXPERT AT THIS---BUT I'D SAY THIS IS A BETTER PLACE TO CROSS-IF WE START FROM HERE WE CAN HEAD ACROSS TOWARDS THAT SNAG-THEN WE'D BE PULLIN' MORE INTO THE CURRENT, AND NOT BROADSIDE LIKE AT OLSEN'S SPOT!

BUT HOW AM I GOIN' TO GET THE LINES FROM THE DONKEY ACROSS THE RIVER?----THAT'S THE FIRST THING I'VE GOT TO FIGURE OUT---I'M BEAT, IF I CAN'T GET A LINE ACROSS TO A STUMP ON THE OTHER SIDE! !?!
Bus Griffiths

THAT BANK LOOKS A BIT HIGH, BUT I'D LIKE TO LOOK AT IT FROM THAT SIDE--- BUT I SURE *CAN'T SWIM* OVER THERE!

I'D BETTER START USING MY HEAD! THAT TIMBER IS ALL DOWN---THAT MEANS THE FALLERS HAD A CROSSING LOG SOMEWHERE UP THE CANYON—

AL WORKS HIS WAY UP ALONG THE CANYON, LOOKING FOR THE POCK-MARKS OF *"CAULKED BOOTS," WHICH WOULD INDICATE THE FALLERS' TRAIL—HE SOON FINDS THE TRAIL AND FOLLOWS IT TO THE CANYON'S EDGE
WOW! NOW *THIS* IS *SOME TRAIL!*

I GUESS I'M LUCKY! HIGH PLACES DON'T REALLY BOTHER ME----BUT A GUY SURE WOULDN'T WANT TO LOOK DOWN AT THAT WATER TOO OFTEN WHEN HE WAS GOING OVER HERE!

SOON AFTER CROSSING TO THE FAR SIDE OF THE CANYON, AL WAS AMONGST THE FELLED AND BUCKED TIMBER—

HE NOTICED THE SHARP SMELL OF THE FRESHLY CUT LOGS, THE SMELL OF THE BRUISED BARK AND SAPWOOD, BUT STRONGEST OF ALL WAS THE PLEASANTLY PUNGENT ODOR OF CRUSHED FIR NEEDLES, WARMED BY THE SUN—

BOY! THAT'S SURE A NICE SMELL! I GUESS LOTS OF GUYS HARDLY NOTICE IT WHEN THEY'RE AROUND IT ALL THE TIME--- BUT YOU SURE NOTICE IT IF YOU'VE BEEN AWAY FROM THE WOODS FOR AWHILE!

AND WHEN YOU'VE BEEN IN CAMP FOR QUITE A SPELL, IT MAKES YOU THINK OF CHRISTMAS, AN' OLD FRIENDS! BUT MOSTLY IT MAKES YOU THINK OF LOGGING, AND I GUESS IT'S A SMELL THAT A LOGGER WOULD NEVER REALLY FORGET!

LOGGING TERMS

*"*CAULKED BOOTS"*— THE BADGE OF A LOGGER'S TRADE—QUITE OFTEN MADE-TO-MEASURE, AND HAND CRAFTED FROM THE FINEST LEATHER—THE SOLES WERE MADE FROM SPECIALLY SELECTED OAK, OR HEMLOCK-TANNED LEATHER INTO WHICH WERE SET THE CAULKS, SHORT SPIKES OF HARD STEEL—THESE GAVE THE LOGGER HIS FOOTHOLD WHEN WORKING ON LOGS, OR OTHER SLIPPERY SURFACES—

WELL, THAT CROSSIN' PLACE LOOKS GOOD FROM THIS SIDE, TOO — THE BANK IS A BIT HIGH, BUT I CAN PULL OUT OF THE WATER ONTO THAT ROCK SHELF, AND THEN MAYBE ROLL A LOG DOWN OFF THE BANK FOR THE DONKEY TO CLIMB ON —

GUESS I'D BETTER HEAD BACK — I'LL PICK UP MY AXE AND BLAZE A LINE FROM THE CROSSIN' SPOT BACK TO THE DONKEY —

DAMMIT… I STILL HAVEN'T FIGURED HOW I'LL GET A LINE FROM THE DONKEY ACROSS THE RIVER!

DRIVING BACK TO CAMP IN THE PICKUP TRUCK, AL RACKS HIS BRAINS TO TRY AND FIGURE A WAY TO GET ONE OF THE DONKEY'S LINES ACROSS THE RIVER —
IT'S NO USE! I'M AFRAID IT'S GOT ME BEAT!

THAT NIGHT AL TOSSES RESTLESSLY —
I SURE HATE TO TELL ART I CAN'T MAKE IT! THAT RIVER MUST BE NEARLY 200 FEET ACROSS, AN' WHEN I CAN'T EVEN FIGURE HOW TO GET A LINE OVER TO THE OTHER SIDE, I MIGHT JUST AS WELL TRY TO THROW THE MACHINE ACROSS!
Bus Griffiths -

SAY! JUST A MINUTE!… I'M GETTIN' AN IDEA!! THROW?… THROW…!! BY GOD, THAT MIGHT BE THE ANSWER!

LOGGING TERMS

*"HANK"— A LOOP-LIKE COIL OF FLEXIBLE MATERIAL OF A SPECIFIC LENGTH— IN THIS CASE MADE IN 50 AND 100 FATHOM COILS- 300-600 FEET —

*"BOOMING GROUND"— THE WATER WHERE THE LOGS ARE SORTED, AND STOWED IN RAFT-LIKE BOOMS FOR WATER TOWAGE —

*"HEAVY PORCELAIN CUPS WITHOUT HANDLES"— THIS TYPE OF CUP WAS QUITE COMMON IN THE COOKHOUSES OF CAMPS DURING THE THIRTIES—

WELL, I'D LIKE TO HAVE RED WITH ME!

THAT'S NO PROBLEM! YOUNG GEORGE HAS BEEN AFTER THE RIGGIN' SLINGER'S JOB ON THE TRACK-SIDE FOR QUITE A SPELL--- HE WANTS TO GET OFF THE COLD-DECKER!

WATCH YOURSELF, MOVIN' THAT MACHINE--- IT AIN'T GONNA *BE EASY!*--- AND YOU BETTER DRIVE THE COLD-DECK CRUMMY---- --AND AL--- *TAKE CARE!*

THANKS, ART- I WON'T TAKE ANY CHANCES-

Driving up the steep logging road, Al noticed that a bank of heavy clouds was moving up fast from the south-east— Even as he watched, the clouds were spilling over the tops of the mountains, and funneling down the canyons and little draws— It reminded Al of water spilling over a giant dam— already the first spots of wind-driven rain were splattering on the wind-shield of the crummy—It looked like the start of a miserable day—

Rain, mixed with sleet, was sweeping across the valley by the time Al drove the crummy into the turnaround, near where the cold-deck donkey was sitting— Wind was already stirring the tops of the trees that lined either side of the gravel road—

LOGGING TERMS

*"**SNAG FALLERS**"— Any snags left standing, after a setting was logged off, that were ten feet, or more, in height, had to be fallen—
This job was usually done by a set of older fallers who were paid by the day instead of the "bushel," or contract rate paid to the regular fallers

SNOOSE THE DONKEY PUNCHER, ENGAGES THE MAINLINE FRICTION LEVER AND THE LINE STARTS TO WIND SLOWLY ONTO THE DRUM— HE JERKS THE THROTTLE OPEN, AND, WITH A THROATY ROAR, THE DONKEY STARTS TO PULL ITSELF TOWARDS THE RIVER —

Logging Terms

*"WINDFALL"— A TREE TOPPLED BY WIND—

*"SHACKLE"—TO FASTEN WITH A SHACKLE, OR CLEVIS—

SHACKLE

*"NOTE"—FALLERS SELDOM WORE THE BONE DRY, OR "TIN COATS," AS THEY WERE TOO STIFF AND RESTRICTED THEIR MOVEMENTS—A LOT OF THEM WORE TWO STANFIELD WOOL SHIRTS---THE UNDER ONE TUCKED IN, THE OUTER WORN OUTSIDE OF TIN PANTS—THE OUTER ONE WAS BOUGHT SEVERAL SIZES TOO LARGE, THEN BOILED IN WATER CONTAINING LINSEED OIL—THE GARMENT WOULD SHRINK AND THICKEN LIKE A BLANKET, AND THE OIL GAVE A LITTLE WATER REPELLENCE— ON A WET DAY THE FALLERS WOULD OCCASIONALLY WRING OUT THE SHIRT TAIL, AND IF IT BECAME SOAKED, WOULD TAKE THE GARMENT OFF AND WRING IT OUT—

Bus Griffiths

WHEN THE DONKEY IS CLOSE TO THE *"TAIL HOLD" TREE, THE LINES ARE SLACKED AND AL PICKS HIS NEXT HOLD—THE MOVING BLOCK AND STRAPS ARE FASTENED ON THIS TREE AND THE LINES STRUNG OUT, READY FOR THE NEXT PULL—AT AL'S SIGNAL, THE DONKEY PUNCHER PULLS THE MACHINE TOWARDS THE NEW TAIL HOLD—
THIS PROCEDURE IS REPEATED A DOZEN OR MORE TIMES DURING THE FORENOON—THE MACHINE MOVES EASILY OVER THE FOREST FLOOR, AND BY MID-DAY THE RIVER IS IN SIGHT—

DURING THE AFTERNOON AL AND HIS CREW GET THE DONKEY MOVED DOWN TO THE RIVER—THE TWO BIG CEDAR LOGS ARE YARDED DOWN, PLACED ONE ON EACH SIDE OF THE DONKEY SLEIGH, AND SECURELY LASHED TO THE RUNNERS, AND BY THEN IT IS QUITTING TIME—

Logging Terms

*"TAIL HOLD"— WHEN MOVING A DONKEY, IF A MOVING LINE IS MADE FAST TO A TREE OR STUMP, OR THREADED THRU A BLOCK FASTENED TO THE OBJECT TO INCREASE POWER, THE TREE OR STUMP IS CALLED THE "TAIL HOLD"—CALLED "TAIL HOLT" BY MANY LOGGERS—

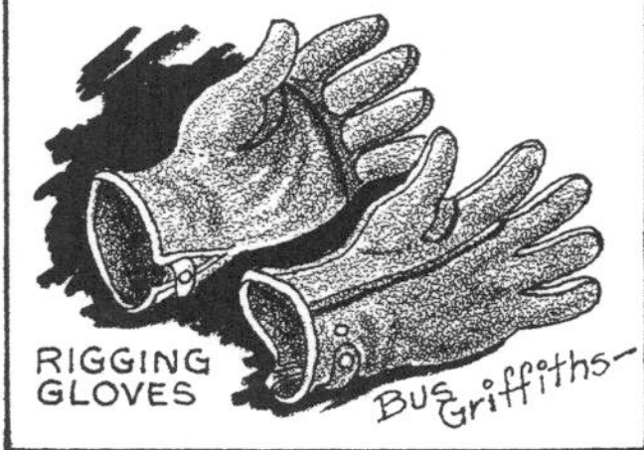

SEVERAL TIMES DURING THE NIGHT AL WAS AWAKENED BY THE FURY OF THE STORM--- HE COULD HEAR THE WIND MOANING AROUND THE EAVES OF THE BUNKHOUSE, AND THE STEADY LASHING OF THE DRIVING RAIN ON THE SHAKE ROOF OF THE BUILDING—

BY MORNING, HOWEVER, THE STORM HAD ABATED AND WAS FOLLOWED BY A SHARP CLEARING, WHICH SHOWED PROMISE OF A GOOD DAY—

RED AND JOHNNY DISAPPEAR INTO THE TIMBER, AND AL STARTS HIS PREPARATIONS TOWARDS GETTING A LINE ACROSS THE RIVER— HE TAKES THE HANK OF LINE, GIVEN TO HIM BY JIM BRADLEY, AND HE AND THE WHISTLE PUNK HEAD UP RIVER FOR THE ROCKY POINT—

Logging Terms

*"PASS ROPE"— A ROPE USED FOR THREADING THE PASSLINE CABLE THRU THE PASSLINE BLOCK IN THE SPAR TREE—GENERALLY ABOUT HALF AN INCH IN DIAMETER, AND 300-400 FEET IN LENGTH—

Bus Griffiths—

AL WHIRLS THE SALMON LINE AROUND HIS HEAD WITH ALL THE POWER IN HIS STRONG YOUNG BODY— THE THIRD TIME THE LINE SWINGS BEHIND HIS HEAD, AND STARTS ON ITS FORWARD ARC, HE RELEASES IT----

THE HEAVY NUT, FASTENED TO THE LINE, CARRIES IT STRAIGHT AND TRUE TO THE FAR BANK OF THE RIVER—

WHILE THE REST OF THE CREW ARE PULLING THE ROPE ACROSS THE RIVER, AL SETS A SMALL BLOCK AND STRAP ON THE BIGHT OF THE STRAWLINE, FASTENING THE END OF THE LINE TO THE LONGEST STRAP ON THE MOVING BLOCK—

AL TIES THE ROPE TO THE GOOSENECK OF THE SMALL BLOCK—

ACROSS THE RIVER—

WE'LL HANG THAT BLOCK ON THIS LITTLE STUMP— WE CAN USE THAT BIG ONE FOR THE MOVING BLOCK!

ONCE THE BIGHT OF THE STRAWLINE WAS ACROSS THE RIVER, IT WAS JUST A MATTER OF MINUTES TO GET THE MOVING BLOCK AND THE DONKEY'S MAINLINE ACROSS— AL HAD HOOKED THE STRAWLINE TO THE EYE OF THE LONGEST STRAP ON THE MOVING BLOCK---THE MAINLINE WAS THREADED THRU THE SHEAVE OF THE BLOCK AND A BIG SHACKLE WAS FASTENED IN THE EYE TO HOLD THE LINE IN PLACE--- *SEE SQUARE No. 4*
RED SIGNALS, AND SNOOSE, THE ENGINEER, GOES AHEAD ON THE STRAWLINE, AND THIS PULLS THE MOVING BLOCK AND THE MAINLINE ACROSS THE RIVER—

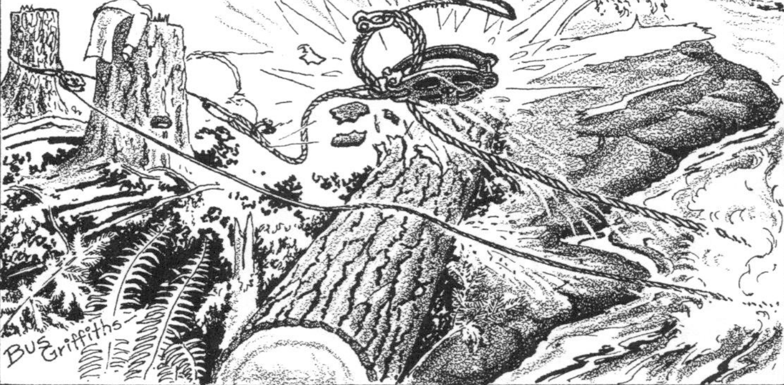

THE STRAPS ON THE MOVING BLOCK ARE SHACKLED TOGETHER AROUND THE BIG FIR STUMP AND THE STRAWLINE HOOKED TO THE EYE OF THE MAINLINE—

RED SIGNALS, THE DONKEY ROARS, AND THE STRAWLINE PULLS THE EYE OF THE MAINLINE BACK ACROSS THE RIVER—

RED HAND-SIGNALLING--"AHEAD ON STRAWLINE"

AL SHACKLES THE MAINLINE TO A STRAP ATTACHED TO A LUG ON THE DONKEY SLEIGH

WE'RE ABOUT READY— DAVE, YOU CLIMB UNDER THE SLEIGH AN' PULL THE HAULBACK LINE OUT THE BACK END--I HUNG A BLOCK ON THE RUNNER--- YOU RUN THE LINE THRU AN' HOOK ONTO THOSE GUYLINES

WILL DO, AL!

Logging Terms

*"GUYLINES"—WIRE ROPE (CABLE) SHACKLED AROUND SPAR TREE NEAR TOP, STRETCHED OUT AND TIGHTENED BY POWER AROUND A NOTCHED STUMP— THE CABLE IS THEN SPIKED TO THE STUMP WITH RAILROAD SPIKES—ANOTHER WRAP IS THROWN ON THE STUMP, TIGHTENED AND SPIKED, THEN A THIRD WRAP—SOMETIMES THE LAST TWO WRAPS ARE THROWN ON AND TIGHTENED, THEN BOTH WRAPS ARE SPIKED AT THE SAME TIME—

THE PURPOSE OF GUYLINES IS TO SUPPORT AND STEADY THE SPAR TREE, AS IT IS UNDER TREMENDOUS STRAIN DURING LOGGING OPERATIONS—

ON A COLD-DECK SPAR, FIVE TOP GUYS ARE USED—

EARLIER IN THE STORY IT WAS MENTIONED THAT THERE WERE TWO SETTINGS ACROSS THE RIVER TO BE COLD-DECKED, HENCE TWO SETS, OR TEN GUYLINES WILL BE NEEDED—

BUS Griffiths—

IN THE MEANTIME—CAMP PUSH, ART DONNEGAN, AND HIS BOSS, JIM BRADLEY, ARE JUST COMING DOWN THRU THE TIMBER, IN SIGHT OF THE RIVER—

UPSTREAM, AND HIDDEN FROM VIEW OF AL AND HIS CREW BY THE BEND IN THE RIVER, A BIG CEDAR, TOPPLED BY THE OVERNIGHT STORM, SWEEPS DOWN ON THE CROSSING DONKEY

IT ROUNDS THE BEND IN THE RIVER—

CARRIED ALONG AT NEAR-EXPRESS-TRAIN SPEED BY THE RAGING WATERS OF THE RAIN-SWOLLEN RIVER, THE BIG CEDAR BEARS DOWN ON THE CROSSING DONKEY AND APPEARS ABOUT TO HIT THE SLEIGH SQUARELY AMIDSHIPS

THE TREE TURNS CROSSWAYS IN THE CURRENT, AND THE TOP JAMS INTO THE BANK OF THE RIVER—

THIS ACTS AS A GIGANTIC *"JILL-POKE", FORCING THE DONKEY SLEIGH TOWARDS THE OPPOSITE BANK OF THE RIVER—

Logging Terms

*"JILL-POKE"— A STIFF-LEG PROP WHICH, WHEN POKED AGAINST A MOVING OBJECT, WILL CAUSE THE OBJECT TO CHANGE DIRECTION— ALSO A DEVICE FOR UNLOADING LOGS—

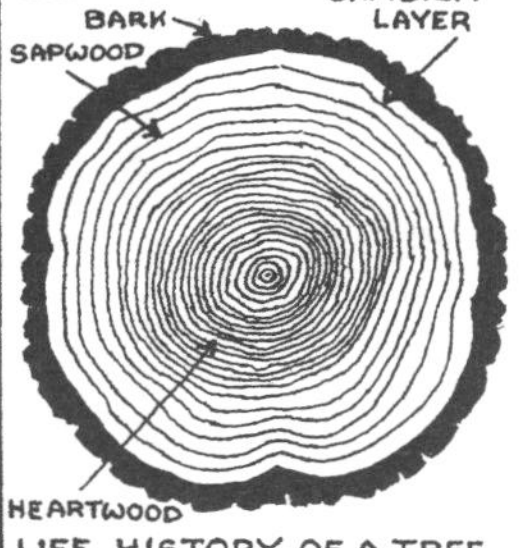

LIFE HISTORY OF A TREE

A SHORT TIME LATER, THE DONKEY NOSES UP ON THE RIVER BANK

THE CROSSING HAS BEEN MADE!

Logging Terms

*"SNOOSE"— (SNUFF) - ALSO CALLED····· "COPENHAGEN", "SCANDIHOOVIAN DYNAMITE", AND "SWEDISH CONDITIONING POWDER"

A FINELY GROUND CHEWING TOBACCO

CONSIDERED VERY ESSENTIAL TO GOOD LOGGING—

HE'S MADE IT, ART! THE LAD'S GOT HER ACROSS!
TOOT-TOOT-TOOT TOOT-TOOT TOOT-TOOT-TOOT TOOT-TOOT TOOT-TOOT-TOOT
YOU'RE DAMNED RIGHT HE HAS! I TOLD YOU HE COULD DO IT!

BY GOD, JIM—WE'VE GOT US A REAL HOOKTENDER! JUST WAIT UNTIL HE LEARNS SOMETHIN'!
HUH ?!?

WELL, BOYS—WE'LL JUST CUT THESE LOGS FREE OF THE RUNNERS, AND THEN---
FOR GOD'S SAKE, MAN, RELAX!--- THEM LOGS AIN'T LIKE GODDAMNED BLUE GROUSE---THEY AIN'T GONNA FLY AWAY----YOU DON'T SEEM TO REALIZE WHAT YOU DONE!

THIS WASN'T LIKE GOIN' ON NO SUNDAY WALK—IT WAS A REAL JOB OF WORK, AN' YOU DONE IT LIKE A REAL PRO!
WELL THANKS, SNOOSE—BUT IF WE HADN'T HAD A COOL CUSTOMER LIKE YOU RUNNIN' MACHINE, AN' A HELLUVA GOOD CREW, IT WOULDN'T HAVE TURNED OUT SO GOOD!

-BUT, YOU'RE RIGHT, SNOOSE---NOTHIN' IS GONNA FLY AWAY! GRAB YOUR *"NOSE-BAGS," BOYS--- WE'LL EAT!
THE OLD MAN AN' ART ARE WAVIN' AN' YELLIN'---WONDER WHAT THEY'RE SAYIN'?
Bus Griffiths

I DON'T GIVE A DAMN WHAT THEY'RE SAYIN'! THEY'RE BOTH GRINNIN' FROM EAR TO EAR----AN' ANY TIME YOU SEE *"THE BRAINS" SMILING, EVERYTHING HAS GOT TO BE ALRIGHT!
*"NOSEBAGS"—LUNCH BUCKETS
*"THE BRAINS"—OWNERS, BOSSES, etc.

It was a cold, bleak-looking morning in mid-November..... The weather had held fairly good thru the fall, without the usual howling south-east gales and heavy rains...... There had been snow near the start of the month, when a sudden storm dumped a foot of the white stuff in the woods and closed down logging operations for two days.... However, the weather moderated, the roads were plowed, and the loggers returned to work............

But this morning the temperature had dipped sharply under a slate-gray sky, and as Al drove the cold-deck crews' crummy up the steep road he cast a speculative glance at the mountains...... He noted that a gray-looking mist was slowly obscuring the higher peaks.......

Logging Terms

*"Sunshine Loggers"—loggers who only like to work in good weather—students who get summer jobs in the woods—

*"Highball"—to do things in a big hurry—

Bus Griffiths—

WHILE YOU CHOP BARK OFF THE DRY SIDE OF THAT FIR SNAG, I'LL MAKE A SHELTER FOR YOU!
GEE- THANKS, AL!

IF YOU'VE GOT ENOUGH FIR BARK FOR YOUR FIRE, YOU CAN STRIP SOME BARK OFF THIS SLABBED-UP CEDAR--CUT IT INTO PIECES ABOUT FOUR FEET LONG!

NOW, KID-- YOU BRING THOSE SLABS OF CEDAR BARK OVER HERE AN' WE'LL LAY THEM ON TOP OF THIS FRAME --

WE'LL LAY A FEW SLABS ON TOP TO BREAK THE JOINTS, AND SHE SHOULD BE WATERTIGHT!
GEE, THAT'S NEAT!

YOU'RE ALL SET FOR THE WINTER, JOEY
THANKS, AL-- YOU KNOW HOW TO DO EVERYTHING---- YOU'RE SURE SMART!
Bob Griffiths

NO, JOEY---- *NOT VERY SMART*, I'M AFRAID---IF I *WAS* SMART, I WOULDN'T BE OUT HERE ON A LOUSY DAY LIKE THIS!

THE SNOW'S GETTIN' HEAVIER···· DAVE MUST BE HAVIN' A HELLUVA TIME CLIMBIN' UP *"THE PILE" TO UNHOOK THE TURNS!

YEAH—WE'LL SEE WHAT SHE LOOKS LIKE AT CHOW TIME! WE MIGHT HAVE TO *"PACK 'ER UP"! I CAN'T REACH IT, JOHNNY··· YOU'LL HAVE TO POKE THE KNOB THRU A BIT FURTHER!

THE CREW TOUGHED IT OUT UNTIL NOONTIME, BUT AFTER LUNCH THE WEATHER DETERIORATED RAPIDLY—

A DRIVING WIND, COUPLED WITH INCREASING SNOW, BROUGHT NEAR-BLIZZARD CONDITIONS—

Logging Terms

*"HANGUP"— WHEN A LOG BECOMES FOULED AGAINST A STUMP, OR OTHER OBJECT—

*"THE PILE"— REFERS TO THE COLD-DECK PILE, WHERE THE LOGS ARE PILED UP AT THE SPAR TREE—

*"PACK 'ER UP"— TO QUIT···· TO GO IN—

Logging Terms

★"TAKE THE LONG SLEEP"— REFERS TO THE WINTER SHUTDOWN—

★"NOTE"— AL IS TIPPING THE GAS BARRELS TO LESSEN THE CHANCE OF MOISTURE SEEPING IN AROUND THE BUNG HOLES, IN THE EVENT OF A LONG SHUTDOWN—

Bus Griffiths—

DO YOU THINK THE BRAINS ARE GONNA LET US USE THE CAT THIS WINTER?
WELL, WHEN I SPOKE TO JIM AND ART, THEY SAID WE COULD HAVE THE CAT, IF THEY WEREN'T USING IT--- BUT DOUG WOULD HAVE TO RUN IT, OF COURSE!
WE WON'T NEED THE CAT FOR AWHILE, ANY-WAY—WE'LL HAVE TO FALL A BUNCH OF THE TIMBER FIRST!
WHAT ABOUT GETTIN' THE TOOLS UP THE SOUND TO JOE'S PLACE?
IN HER LAST LETTER, DEBBIE SAID THAT HER BROTHER, HARV, WOULD BE COMIN' DOWN TO THE VILLAGE FOR SUPPLIES- HE SHOULD BE DOWN IN HIS BOAT ANY DAY, NOW!
THAT MIGHT WORK OUT PRETTY GOOD, IF THEY DO DECIDE TO SHUT DOWN THE CAMP!
YEAH—WE COULD GET OUR GEAR, ROUND UP SOME FALLIN' TOOLS AND GO UP THE SOUND WITH HARV, IN HIS BOAT!
BOY! IT'S SURE A BITCH OF A DAY— IT'LL BE GOOD TO GET IN THE CRUMMY! I'LL HELP YOU CLEAN THE SNOW OFF THE WINDSHIELD!
YOU RIDE DOWN WITH ME, SNOOSE! IT'LL BE A LOT NICER, SITTIN' IN THE WARM CAB THAN RIDIN' IN THE BACK OF THAT GODDAMN DRAFTY OL' CRUMMY!
YOU'RE NOT FOOLIN' ME, ART! YOU'RE SCARED YOU MIGHT GET STUCK IN THE SNOW, AN' YOU FIGURE I'LL BE HANDY FOR WIELDING THE *"MUCK STICK" TO DIG YOU OUT!
YOU GO FIRST, ART! WE'LL FOLLOW IN THE CRUMMY---IF YOU GET STUCK, WE CAN ALL DIG!
YOU GUYS JUST WANT ME TO BREAK TRAIL FOR YOU! I CAN DO IT, TOO! C'MON, SNOOSE---- WE'LL SHOW THOSE BASTARDS!
*"MUCK STICK" — SHOVEL
BUS Griffiths

NEXT MORNING

Logging Terms

*"DEADHEAD"—A LOG THAT FLOATS ON END OR NEARLY SO—
*"SINKERS"—LOGS THAT FLOAT BELOW THE SURFACE, OR SINK TO THE BOTTOM—
*"BOOMSTICKS"—GENERALLY 66 ft. LONG, WITH A HOLE BORED IN EACH END—COUPLED TOGETHER WITH CHAINS THEY FORM A PEN INTO WHICH LOGS ARE STOWED FOR WATER TOWAGE—
*"SECTION"—REFERS TO A SECTION OF LOGS, 66 ft. LONG BY 66 ft. WIDE—WHEN THE RAFT-LIKE BOOMS ARE MADE UP A 66 ft. LOG, CALLED A "SWIFTER", IS PULLED ACROSS THE BOOM AT EACH JOINT OF THE SIDE STICKS AND FASTENED BY CHAIN TO THE CHAIN THAT COUPLES THE STICKS-THIS STOPS THE BOOM FROM BAGGING OUT AND MAKES A "SECTION" OF LOGS

SIDE STICKS
HEAD STICK · HEAD RIDER · SWIFTER · SECTION · SWIFTER · SECTION · SWIFTER · TAIL RIDER · TAIL STICK

BIRD'S-EYE VIEW OF 4-SECTION BOOM

OKAY, YOU DUDES— THROW YOUR GEAR IN JIM'S PICKUP AN' CLIMB ABOARD THE CRUMMIES, AN' LET'S GET THE HELL OUT OF HERE-- WE AIN'T GOT ALL DAY!
WHO'S HE GONNA HOLLER AT, AFTER WE'VE ALL GONE TO TOWN?
HE'LL LIKELY BE OUT IN THE WOODS SCREAMING AT THE RAVENS, OR HOWLING WITH THE WOLVES--- JUST TO KEEP HIS VOICE IN SHAPE!

KINDA SORRY TO SEE 'EM GOIN' ART?
WHO---ME?---- YEAH---I GUESS I AM--FOR THE MOST PART THEY'RE A PRETTY GOOD LOT-- A FEW KNOTHEADS, BUT MOSTLY GOOD--- HOPE WE SEE MOST OF 'EM BACK!

LATER—AT THE BOAT—
YOU GUYS SURE LOOK LIKE A BUNCH OF DUDES---- ALL DRESSED UP IN THEM "STORE-BOUGHT" CLOTHES!

YOU'RE JUST JEALOUS 'CAUSE YOU AIN'T GOIN' TO TOWN WITH US!---- BETTER COME ALONG, AL!

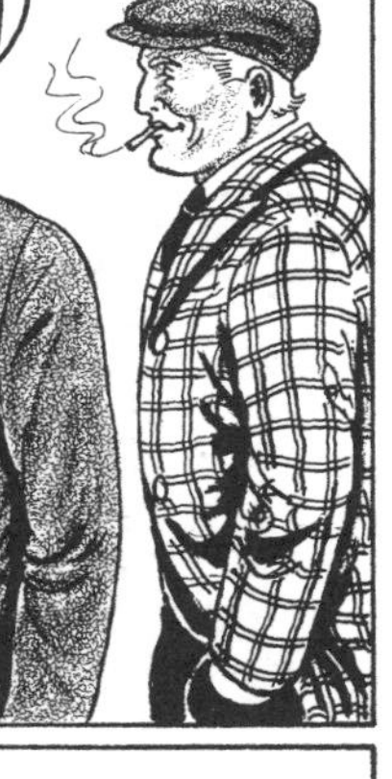

YOU'RE FORGETTIN' THAT AL IS GOIN' UP THE SOUND TO SEE HIS LITTLE GAL!---- HE SAYS HE'S GOIN' UP TO DO SOME LOGGIN', BUT I WASN'T BORN YESTERDAY!

TAKE CARE, YOU GUYS, AN' HAVE A REAL GOOD HOLIDAY! I WANT ALL OF YOU BACK ON MY CREW IN THE SPRING--- AN' YOU BETTER BE IN GOOD SHAPE---WE GOT LOGS TO TAKE OUT!
YOU'RE BEGINNIN' TO SOUND JUST LIKE ART!
WELL, SO LONG, AL- GOOD LUCK WITH YOUR LOGGIN' AND LOVIN' UP THE SOUND—KEEP AN EYE ON HIM, RED!
I'LL DO THAT, SNOOSE!

I ALMOST WISH WE WERE GOIN' WITH 'EM, RED! THEY'RE A GOOD BUNCH OF GUYS!
YEAH—IT'D BE FUN, BUT I'M LOOKIN' FORWARD TO GOIN' UP THE SOUND, AN' LOGGIN' ON OUR OWN---AN' I KNOW YOU ARE, TOO!
BUS Griffiths

I SURE WOULDN'T WANT TO LIVE AN' WORK IN THE CITY··· WOULD YOU, AL ?
NO WAY, RED ! OH, IT'S OKAY IF YOU LIKE THAT KIND OF LIFE, BUT IT'S NOT FOR ME!
IT'S REALLY TOUGH IN TOWN DURING THESE HARD TIMES! THERE'S NOT MUCH WORK, AND EVERY-THING YOU DO COSTS MONEY!
YEAH – IN THE STICKS A GUY CAN FIND WORK OF SOME KIND···YOU MIGHT NOT MAKE A LOT OF MONEY, BUT AT LEAST YOU'LL EAT FAIRLY REGULAR!
THAT'S TRUE···AN' IF A GUY'S GOT A FAMILY HE CAN GROW A GARDEN, AN' HAVE A FEW CHICKENS··· AND MAYBE SOME PIGS AND AN OLD COW!
YEAH– AN' YOU CAN ALWAYS DIG A FEW CLAMS, AND GET OYSTERS···
SURE··· AND THERE'S HUNTIN' AN' FISHIN'····· THEY CAN HAVE THE CITIES···
BUT IT'S KINDA NICE TO GO IN ONCE IN AWHILE FOR A SPREE···· AND THEN GET THE HELL OUT···FAST! OH··HERE COMES OL' "SMILEY"!
BUS GRIFFITHS
I PICKED UP THE CAMP MAIL, SO I'LL HAVE TO GO ABOARD THE BOAT···THERE'S MAIL FOR SOME OF THE BOYS···OH, I THINK THERE'S A LETTER FOR YOU, AL··· WANT TO GUESS WHO IT'S FROM?
THANKS, ART– JUST GIVE ME THE LETTER!
WHAT DOES SHE SAY, AL?
AW– HAVE A HEART, RED– GIVE ME A CHANCE TO READ IT!
We were
but Dad just got
and Harvey will be going down to
the village for supplies this Thursday.
He also has to pick up some gear and
stuff for his boat, so he will be
there most of the morning.
I noticed, the other day, that there is
fresh snow quite low down on the
high mountain across the river from
us, and I'm hoping that you will be
snowed out at the camp. Then you
can come up with Harv and start
work on Dad's timber.
Oh, I'm so excited, and the days
seem so long. I want so much to
see you again, my darling, and
hold you close, the way I did that
day on the trail to the beach, that
seems so very long ago, now.
My thoughts are always with you

I'VE ALWAYS GOT TO BE THE MULE··· SUPPLYIN' THE MOTIVE POWER!

AW, QUIT YOUR BELLY-ACHIN' AN' TURN THE HANDLE··· I WANT THESE AXES SHARP ENOUGH TO SHAVE WITH·· WE'VE GOT SOME MIGHTY BIG UNDER-CUTS TO CHOP, IN THAT TIMBER UP THE SOUND!

YOU DID QUITE A BIT OF FALLIN', DIDN'T YOU, AL?

I FELL TIMBER FOR TWO YEARS, BUT I HAD A REAL GOOD TEACHER···A CRANKY OLD SWEDE, ERIK ERIKSON! A NICE OLD GUY, BUT SURE CRABBY···AND HE COULD SURE FALL TIMBER!

Logging Terms

★"FILING SHACK"— WHERE THE SAW FILER SHARPENS THE TOOLS USED BY THE FALLERS AND BUCKERS—

★"DRAGGING MY FEET"— WHEN A MAN IS HEAVY ON THE SAW, HE IS ACCUSED OF "DRAGGING HIS FEET", OR "RIDING THE SAW"—

AN HOUR'S DRIVE BROUGHT THE MEN DOWN TO THE WHARF AT THE VILLAGE— A GLANCE OVER THE CLUSTER OF BOATS MOORED AT THE FLOATS REVEALED TWO CABIN CRUISERS, A SMALL TUG, & A HALF DOZEN BOATS EQUIPPED WITH THE DRUMS & STERN ROLLERS OF GILLNETTERS— NOWHERE, COULD THEY SEE A BOAT WITH THE TALL, STATELY POLES OF A SALMON TROLLER~

Logging Terms

★ "WIDOW MAKER"

A LOOSE LIMB OR CHUNK, HANGING IN A TREE, WAITING TO FALL ON SOME UNWARY LOGGER'S HEAD— ALSO, ANY DANGEROUS TREE, OR SNAG, IS A POTENTIAL "WIDOW MAKER"

A WHITE-HULLED BOAT ROUNDS THE POINT & ENTERS THE BAY—FROM HER BOWS, TWO POLES JUT OUT AT A RAKISH ANGLE, & AMIDSHIPS, ON EITHER SIDE OF HER MAST, SLENDER POLES REACH SKYWARD··· SHE IS A SALMON TROLLER!

ANYONE WHO HAS WORKED THE SEA, & LOVES BOATS, HAS TO BE THRILLED BY THE SIGHT OF A TRIM, WELL-RIGGED TROLLER UNDER-WAY—TO SEE HER TALL POLES GLEAMING IN THE SUN, & WATCH THE LOVELY LIFT & DIP OF HER BOW AS SHE RISES & FALLS IN A SLIGHT SWELL, SHE IS INDEED AN INSPIRING SIGHT!

"★" YOU WILL NOTICE, IN THE PICTURE, THAT THE BOAT HAS NO POWER GURDIES, AS DO MODERN DAY TROLLERS—AT THIS POINT IN TIME, THE LINES WERE PULLED BY HAND···· COTTON LINES, WEIGHTED DOWN, AT TIMES, WITH AS MUCH AS 20-25 LBS. OF LEAD, IN GRADUATING SIZES, THE HEAVIEST LEADS BEING CLOSEST TO THE SURFACE—

IT WAS NEAR THE END OF THE THIRTIES BEFORE WIRE LINES AND POWER GURDIES CAME TO THE FORE, AND SOME TROLLERS WERE STILL PULLING THE LINES BY HAND WELL INTO THE FORTIES—

Bus Griffiths—

EARLY AFTERNOON FINDS OUR FRIENDS ABOARD HARV'S FISHBOAT, HEADING UP THE SOUND— EACH TURN OF THE MOTOR'S BIG FLYWHEEL BRINGS THEM EVER CLOSER TO THEIR DESTINATION, AND NEW ADVENTURES—

MEANWHILE, FURTHER UP THE SOUND, A PRETTY GIRL SITS BESIDE A BIG FIR TREE, HIGH ATOP A ROCKY BLUFF OVERLOOKING THE SEA—SHE SPEAKS TO HER COMPANION, A BIG SHAGGY DOG, AS SHE ANXIOUSLY SCANS THE SPARKLING WATER OF THE SOUND FOR SIGNS OF HER BROTHER'S TROLLER—

ABOARD THE FISHBOAT

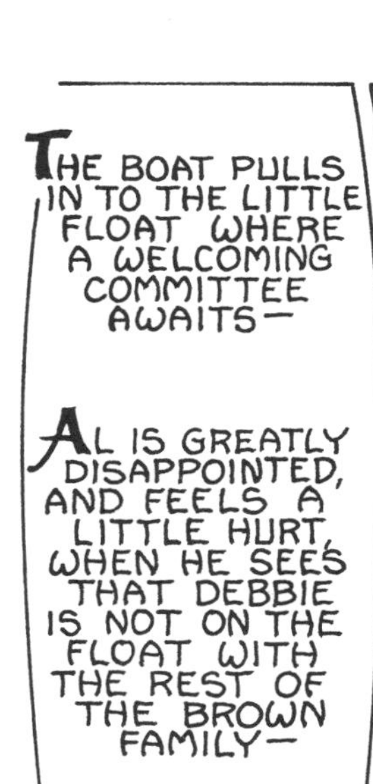
THE BOAT PULLS IN TO THE LITTLE FLOAT WHERE A WELCOMING COMMITTEE AWAITS—
AL IS GREATLY DISAPPOINTED, AND FEELS A LITTLE HURT, WHEN HE SEES THAT DEBBIE IS NOT ON THE FLOAT WITH THE REST OF THE BROWN FAMILY—

IT'S GOOD TO SEE YOU LADS AGAIN! YOU CERTAINLY LOOK FIT AS FIDDLES!
I'M SO GLAD YOU'VE COME!
THANKS···IT'S SURE NICE SEEIN' YOU FOLKS, AN' YOU'RE BOTH LOOKIN' REAL GOOD!
HOWDY, FOLKS!

THIS IS YOUNG BRUCIE···OUR BABY!
AW-W MOM!

HI, BRUCE— YOU DON'T LOOK LIKE A BABY TO ME! I'M COUNTING ON YOUR HELP WHEN WE START LOGGIN'! ···BUT···ER··A··· WHERE'S DEBBIE?
I BELIEVE SHE WENT FOR A WALK WITH THE DOG!
I KNOW WHERE SHE WENT···I'LL SHOW YOU!

IF YOU FOLKS WILL EXCUSE ME···I'D LIKE TO····
OF COURSE, AL!
GO AHEAD, SON!
I'LL HELP HARV WITH THE TOOLS AN' STUFF, AL!
COME ON— I'LL SHOW YOU!

SIS HAS GOT A PLACE SHE GOES WHEN SHE WANTS TO BE ALONE—YOU GO UP THIS TRAIL AND YOU'LL FIND HER!
THANKS, PARTNER!

I WONDER WHY DEB WASN'T THERE TO MEET ME?··· AND AFTER THE LETTERS SAYIN' SHE COULDN'T WAIT TO SEE ME AGAIN! I GUESS A GUY *NEVER CAN* UNDER-STAND WOMEN!
Bus Griffiths—

WE TURN TO DEBBIE, COMING DOWN THE TRAIL TOWARDS AL—
MAYBE I SHOULD HAVE BEEN THERE TO MEET AL---- MAYBE HE WON'T COME LOOKING FOR ME!

A CURVE IN THE TRAIL!
OH, SHEP--IT'S HIM! NOW DON'T YOU DARE GROWL OR BARK AT HIM!--THAT'S OUR LOGGER MAN!

AL! OH, AL!
DEBBIE!

DEAR LITTLE DEBBIE!

HEY! NOW, POOCH!... HEY!
BUS Griffiths-

WELL!---- I HADN'T FIGURED ON ANYTHIN' LIKE THIS!!--- BEIN' LOVED TO DEATH BY AN OLD FUZZY-FACED DOG!
HA! HA! HA! HE'S JUST SHOWING HIS APPROVAL--- HE LIKES YOU!

HE'S QUITE A DOG---BUT I DON'T REMEMBER YOU HAVIN' HIM WHEN I WAS HERE BEFORE!
HE WAS ON THE BOAT WITH THE BOYS, WHEN THEY WERE FISHING, THIS SUMMER!

I'M GLAD HE WASN'T RUNNIN' INTERFERENCE FOR YOU, THE LAST TIME I WAS HERE!
OH, AL---I'M SO SORRY--IT HASN'T TURNED OUT--THE WAY---I--HOPED--IT WOULD!

DEAR LITTLE DEBBIE!---HEY! NO CRYIN', NOW!---I'VE WAITED SO LONG FOR THIS DAY---DON'T CRY AND SPOIL IT ALL!
OH, AL! MY LOGGER MAN!

BUT WHY DIDN'T YOU MEET ME AT THE FLOAT?
I WANTED IT TO BE LIKE IT WAS WHEN WE SAID "GOODBYE"--JUST THE TWO OF US! YOU WOULDN'T HAVE KISSED ME LIKE THAT, WITH MY MOM AND DAD STANDING THERE---WOULD YOU?
BUS GRIFFITHS

WOMEN!---AND YOU NEVER KNOW---I MIGHT JUST HAVE GRABBED YOU AN' KISSED YOU IN FRONT OF EVERYBODY!
OH, YOU!

Logging Terms— *"JUNK"— CLOTHES, BAGGAGE, etc.

THE NEXT TWO PAGES BREAK THE NARRATIVE OF THE STORY & WILL BE MORE OR LESS TECHNICAL—

IN THEM I WILL DESCRIBE THE WORK OF THOSE HIGHLY SKILLED MEN, ENGAGED IN THAT MOST DANGEROUS OF JOBS IN THE COAST WOODS, HAND FALLING & BUCKING TIMBER—

TIM-BER-R-R!! THE CRY ECHOES THRU THE WOODS··· THE SOUND OF STEEL WEDGES, DRIVEN WITH A HEAVY HAMMER, FOLLOWED BY THE *"ZIP, ZIP"* OF A FAST-PULLED FALLING SAW!··· THEN THE CRY GOES UP AGAIN, *"TIM-BER···BACK IN THE WOODS"!* THE SOUND ROLLS THRU THE WOODS LIKE THE DEEP HOWL OF A WOLF—

WOOD FIBRES TEAR APART, & AN ALMOST HUMAN CRY SEEMS TO COME FROM THE DOOMED TREE AS IT LEAPS FROM THE STUMP & CRASHES TO EARTH, ENDING A LIFE-SPAN OF HUNDREDS OF YEARS!···· *THE HAND FALLERS ARE AT WORK!*

A "SET" OF HAND FALLERS USUALLY CONSISTED OF A HEAD & SECOND FALLER, & ONE OR TWO BUCKERS—THE HEAD FALLER WAS THE BOSS—HE PICKED THE "LEAN" OF THE TREES, & WAS RESPONSIBLE FOR LAYING OUT THE TIMBER—WHEN MOVING FROM TREE TO TREE, HE PACKED THE SAW, OIL BOTTLE, HIS AXE, & ONE SPRINGBOARD—THE SECOND FALLER TOOK THE HAMMER, WEDGES, HIS AXE, & THE OTHER SPRINGBOARD—IF THE HEAD FALLER WAS A MISERABLE CHARACTER, HIS PARTNER WOULD BE PACKING BOTH BOARDS—WHEN APPROACHING A TREE, THE HEAD FALLER WOULD GLANCE UP & PICK THE "LEAN" OF THE TREE, THEN DROP THE SAW AT WHAT WOULD BE THE BACK OF THE TREE—HIS PARTNER SOON LEARNED TO WATCH FOR THIS, & KNEW, WITHOUT ASKING, WHICH WAY THE TREE WOULD FALL—ONCE THE LINE OF FALL OF A BIG TREE WAS PICKED, THE HEAD FALLER WALKED OVER THE GROUND TO MAKE SURE THERE WAS A "GOOD LAY" FOR THE TREE—IF THE GROUND WAS TOO UNEVEN, SOME JUNK TREES WOULD BE FALLEN TO LEVEL OFF THE LAY, BUT IF TOO BAD, ANOTHER LAY WOULD BE PICKED—THE TIMBER WAS USUALLY FALLEN "IN LEAD" WITH THE SPAR TREE, WHEN FALLING A SETTING FOR "HIGH-LEAD" LOGGING—IN SIMPLE TERMS: TOWARDS, OR AWAY FROM THE SPAR—TREES WITH A BAD SIDE LEAN WERE FALLEN ON TOP OF THE OTHER LOGS, SO THEY COULD BE EASILY SWUNG INTO LEAD WITH THE DONKEY, WHEN YARDING THE SETTING—

EACH FALLER HAD A DOUBLE-BITTED FALLING AXE, SPECIALLY MADE FOR THE JOB—FOUR & A HALF POUND HEADS, WITH NARROW, THIN BLADES, HUNG ON 42 INCH HANDLES OF SELECT HICKORY— ONE BIT WAS HONED TO NEAR RAZOR SHARPNESS FOR CHOPPING SPRINGBOARD NOTCHES & UNDERCUTS, THE OTHER BIT SHARP, BUT STUBBED OFF FOR CHOPPING BRUSH & LIMBS—

THE SAW USED WAS A THIN-BLADED, TWO-MAN CROSSCUT, GENERALLY SEVEN OR EIGHT FEET IN LENGTH—

WHAT WE USED TO DO, WHEN WORKING IN BIG STUFF, WAS START THE BACK CUT WITH A SEVEN-FOOT SAW, & WHEN WE RAN SHORT OF STROKE WE'D USE A TEN OR TWELVE-FOOT SAW TO FINISH THE CUT—EVEN THEN, OCCASIONALLY, WE'D HAVE TO CHOP SIDE NOTCHES IN A TREE TO GET ENOUGH STROKE TO FINISH THE CUT—

THIS WAS ESPECIALLY TRUE WHEN WORKING IN THE BIG CEDARS IN THE FRASER VALLEY—

SPRINGBOARDS WERE USED TO GET ABOVE THE SWELL OF THE BUTT, & THEY ALSO MADE A NICE LEVEL PLACE ON WHICH TO STAND WHILE CHOPPING & SAWING—

THE BOARDS WERE GENERALLY 4 FEET LONG, 6 INCHES WIDE, & 2 INCHES THICK, & YELLOW CEDAR WAS A FAVORED WOOD—

A STEEL PLATE, WITH A LIP ON THE TOP SIDE OF THE ROUNDED END, WAS BOLTED TO THE SHAPED END OF THE BOARD WITH 3 BOLTS—THIS END OF THE BOARD WAS INSERTED IN THE SPRINGBOARD HOLES CHOPPED IN THE BUTT OF THE TREE BY THE FALLERS—

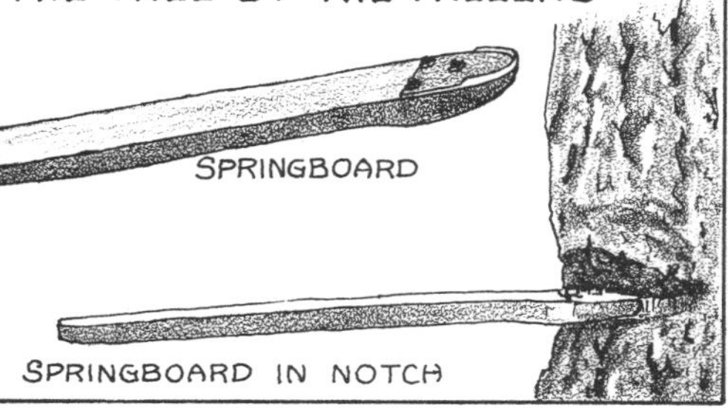

SOMETIMES A LITTLE PERSUASION WAS NEEDED TO MAKE A TREE FALL IN THE DESIRED SPOT—AN 8-12 lb. HAMMER & STEEL WEDGES WERE THE PERSUADERS—EVERY SET OF FALLERS HAD THREE OR FOUR LONG, 8-10 lb. FALLING WEDGES—WHEN WORKING IN REAL BIG, HEAVY TIMBER, A STEEL PLATE ABOUT 10"X12" WAS ALSO IN THE WEDGE SACK—WHEN THE SAW CUT STARTED TO BIND, THE THIN EDGE OF A WEDGE WAS DRIVEN INTO THE CUT BEHIND THE SAW—THE GENERAL PRACTICE WAS TO PLACE TWO WEDGES, SPACED A LITTLE APART IN THE CUT, & DRIVE THEM IN UNTIL THE TREE STARTED TO LIFT—IF THE TREE WAS VERY "HEAVY" THE STEEL PLATE WOULD BE INSERTED IN THE CUT, A LITTLE OIL SHAKEN ON IT FROM THE OIL BOTTLE, & A WEDGE PLACED ON TOP & DRIVEN IN—THIS USUALLY HAD THE DESIRED EFFECT—

THE WEDGE SACK WAS MADE FROM A WIDE FEED SACK, CUT AS IN THE SKETCH—THE TWO SIDE STRIPS WERE TIED TOGETHER WITH A REEF KNOT, & MADE THE STRAP OF THE SACK—THE SLEDGE HAMMER HEAD WAS "FIGURE-EIGHTED" AROUND THE STRAP, & WHEN THE SECOND FALLER THREW THE WEDGE SACK OVER HIS SHOULDER, THE HAMMER HANDLE WOULD LAY UNDER HIS ARMPIT, LEAVING THAT HAND FREE TO PACK ANOTHER TOOL—

THE OIL BOTTLE WAS QUITE OFTEN A JAPANESE SAKI BOTTLE WITH A LARGE, SHARP HOOK LASHED TO THE NECK WITH TARRED HEMP—THE CORK WAS WHITTLED FROM A PIECE OF LIVE FIR BARK & HAD TWO OR THREE GROOVES CUT DOWN THE SIDES SO OIL COULD BE SHAKEN OUT—

THE OIL, USUALLY STOVE OR COAL OIL, WAS USED TO LUBRICATE THE SAW, & CUT THE PITCH THAT WAS OFTEN ENCOUNTERED IN THE LARGE FIRS—

WE FOUND THAT IN CERTAIN TYPES OF PITCH, WATER WORKED BETTER THAN OIL—AN AXE WOULD BE DRIVEN INTO THE TRUNK OF THE TREE ABOVE THE SAW CUT, & A CANVAS WATER BAG HUNG ON THE AXE SO THAT A THIN STREAM OF WATER RAN ON THE SAW BLADE AS IT WAS PULLED THRU THE CUT—

BUS GRIFFITHS—

THE BUCKER!........ HIS JOB WAS TO CUT THE FELLED TIMBER INTO LOG LENGTHS, WITH AN EYE TO SCALE & GRADE— IT WAS A HARD, DANGEROUS JOB, REQUIRING MEN OF MUCH PATIENCE & ENDURANCE— THE BUCKER WAS OFTEN REFERRED TO AS THE "MONEY MAKER" OF THE FALLING CREW— BY CAREFUL & SKILLFUL CUTTING, HE'D MAKE HIS CUTS TO GET THE MOST POSSIBLE "BOARD FEET" OF SCALE OUT OF EVERY TREE— THIS WAS VERY IMPORTANT, AS THEY WORKED ON "CONTRACT" & WERE PAID A CERTAIN AMOUNT PER THOUSAND BOARD FEET OF TIMBER, FELLED & BUCKED, AS SCALED BY A CAMP LOG SCALER— THE BUCKER ALSO MADE HIS CUTS WITH AN EYE TO GETTING THE BEST GRADE OF LOGS FROM EACH TREE— SO HE NOT ONLY MADE MONEY FOR HIMSELF & HIS PARTNERS BY BUCKING FOR SCALE, BUT WAS ALSO AN ASSET TO HIS EMPLOYERS BY CUTTING THE TREES FOR GRADE & MAKING GOOD LOGS—

THE SAW USED BY THE BUCKER WAS SPECIALLY MADE FOR THE JOB··· A LOT DEEPER FROM BACK TO POINT OF TOOTH THAN A FALLING SAW, & SEVEN OR EIGHT FOOT LENGTHS WERE MOST COMMONLY USED—
IT WAS A TWO-MAN CROSSCUT, BUT ON THE PACIFIC COAST OF BRITISH COLUMBIA, THE BUCKER NEVER HAD A PARTNER ON THE SAW— IF THERE WERE TWO BUCKERS IN A "SET," EACH ONE HAD HIS OWN SAW & WORKED ALONE—

AN AXE SIMILAR TO THE FALLERS' AXES WAS USED BY THE BUCKER, THE DIFFERENCE BEING IN THE HANDLE— THE HANDLE ON THE BUCKER'S AXE WAS SHORTER, RARELY LONGER THAN 3 FEET, & HAD A SERIES OF NOTCHES CUT IN ONE SIDE, ABOUT A FOOT FROM THE END— THE NOTCHES WERE PUT THERE BY THE BUCKER WITH EITHER A KNIFE OR A FILE, & WERE USED FOR "UNDERBUCKING"—

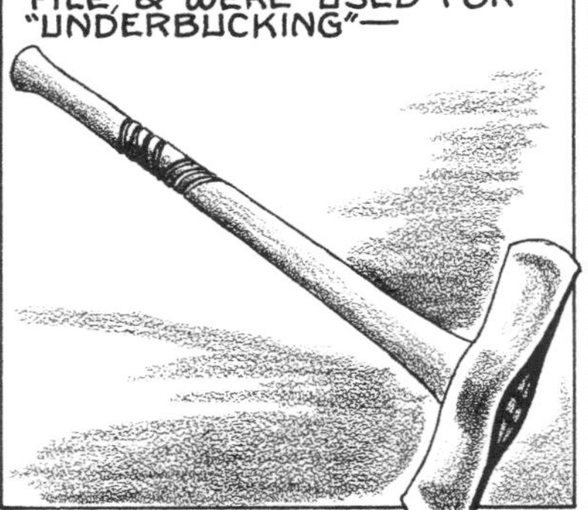

WHEN THE TREE WAS CLEAR OF THE GROUND WHERE A CUT WAS TO BE MADE, THE CUT WOULD USUALLY BE STARTED OR FINISHED BY "UNDERBUCKING"— IF THE CUT WAS GOING TO BIND, THE TOP CUT WOULD BE MADE FIRST, & FINISHED BY "UNDERBUCKING"— THE BUCKER WILL DECIDE WHICH LOG WILL DROP THE LEAST WHEN THE CUT IS MADE, & DRIVE HIS AXE INTO THE LOG SO THAT THE HANDLE WILL BE MORE OR LESS PARALLEL TO THE LAY OF THE LOG— HE SHAKES A LITTLE OIL ON THE HANDLE IN THE AREA OF THE NOTCHES— TURNING HIS SAW UPSIDE DOWN HE WILL PLACE THE BACK IN ONE OF THE NOTCHES, & USING THE AXE HANDLE AS A GUIDE, & FOR LEVERAGE, WILL SAW FROM THE BOTTOM OF THE LOG UP TO MEET HIS TOP CUT— WHEN THE CUT IS FINISHED IT WILL BE ABOUT A HALF INCH CLOSER TO THE AXEHEAD THAN THE TOP CUT, THUS ALLOWING THE OTHER LOG TO DROP CLEAR— A LAYING TREE, WITH THE TOP HANGING, WOULD SPLIT IF THE CUT WAS STARTED FROM THE TOP SIDE OF THE LOG, SO THE BUCKER WOULD "UNDERBUCK" FIRST TO RELIEVE THE STRESS— THEN HE WOULD BUCK DOWN IN THE NORMAL MANNER TO FINISH THE CUT—

THE BUCKER HAD A SLEDGE HAMMER & A WEDGE SACK CONTAINING THREE OR FOUR BUCKING WEDGES— THESE DIFFERED FROM THE FALLERS' WEDGES, AS YOU WILL SEE IN THE SKETCHES— THEY WERE NOT AS LONG AS FALLING WEDGES, & THE BOTTOM WAS MORE FAN SHAPED— THE MAIN REASON FOR THIS FAN SHAPE WAS BECAUSE IT MADE THE WEDGES MORE EFFECTIVE WHEN USED AS "HANGING WEDGES"··· THAT IS, DRIVEN INTO THE LOG CROSSWAYS TO THE CUT— THIS WAS DONE TO STOP A CUT FROM OPENING TOO QUICKLY & SPLITTING THE LOG, & ALSO TO STOP THE LOG FROM ROLLING ON THE BUCKER WHEN THE CUT WAS FINISHED!

BUCKING WEDGE

FALLING WEDGE

THE BUCKER HAD A MEASURING STICK WHICH HE USED FOR MEASURING THE LOGS BEFORE MAKING HIS CUTS— THE STICK WAS 8 FEET LONG & MARKED OFF EVERY 2 FEET—

WHEN THE BUCKER MEASURED HIS LOGS, HE ALLOWED AT LEAST AN EXTRA 6 INCHES FOR WHAT WAS CALLED "SNIPE"— THIS WAS TO COMPENSATE FOR CROOKED CUTS—

Bus Griffiths

★"NOTE"—

THE AXEHEAD IS PLACED AGAINST THE BACK OF THE UNDERCUT, AS IN THE SKETCH— HOLDING THE AXEHEAD IN PLACE, AL SIGHTS ALONG THE AXE HANDLE TO DETERMINE THE LINE OF THE TREE'S FALL—THIS IS CALLED "GUNNING"—IF THE LINE OF FALL ISN'T ON TARGET, THE FALLERS WILL CHANGE THE ANGLE OF THE BACK OF THE UNDERCUT WITH THEIR AXES, UNTIL THE LINE OF FALL IS ON TARGET— (*see dotted lines*)

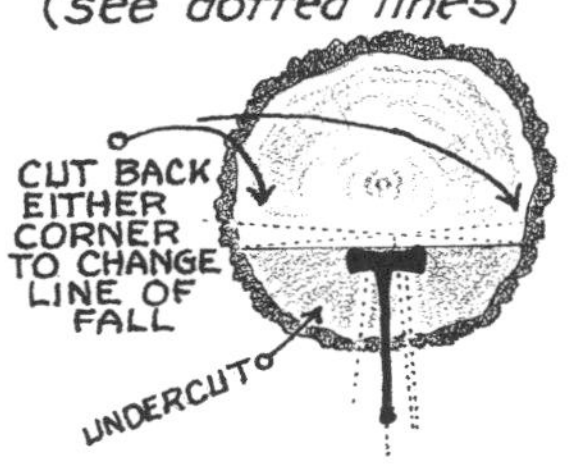

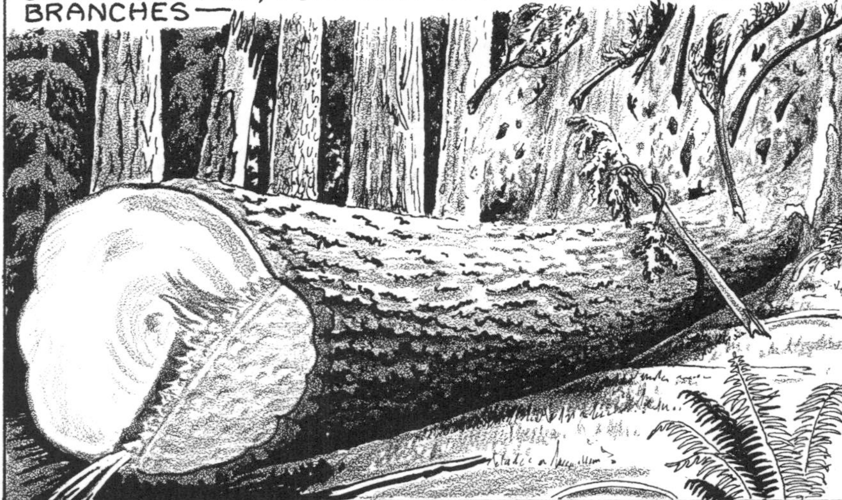

"★" LOGGERS' TERM MEANING··· THE TREE LAYS OUT, UNBROKEN, FOR ITS FULL LENGTH····

AL IS BUCKING THE BUTT LOG OFF THE BIG CEDAR, UNDER THE WATCHFUL EYES OF JOE BROWN AND YOUNG BRUCE—THE CUT IS OPENING, AND AL HAS DRIVEN ★"HANGING WEDGES" IN ON EACH SIDE OF THE LOG TO STOP THE CUT FROM OPENING TOO QUICKLY— HE THEN PICKS UP A STRAIGHT, FINGER-SIZED LIMB, AND CHOPS OFF A FOOT-LONG PIECE— HE DROPS IT IN THE SAW CUT, ABOVE THE BACK OF THE SAW—

HOW COME YOU CUT HALF WAY THRU THE BUTT LOG, PUT WEDGES IN IT, AND NOW YOU'RE LEAVIN' IT?
WELL, I CAN'T GET UNDERNEATH TO UNDERBUCK, AN' IF I SAWED ANY MORE, THE LOGS WOULD SPLIT!

WHY DIDN'T YOU MAKE ONE OF THE OTHER CUTS FIRST?
YOU HAVE TO SIZE UP THE WHOLE TREE, HARV, AN' FIGURE WHAT IS GONNA HAPPEN WHEN YOU TAKE OFF EACH CUT!

THE WAY THE TREE LAYS, THERE'S A SLIGHT HUMP IN IT—THAT MEANS THE BUTT CUT WILL BE OPENING··· BUT LIKE I SAID, YOU CAN'T CUT IT MORE THAN HALF WAY, OR IT WILL SPLIT—BUT IF YOU MAKE ONE OF THE OTHER CUTS FIRST, THE BUTT CHUNK WILL DROP, AND THAT WILL MAKE THE BUTT CUT BIND!

THE BUTT CUT IS THE BIGGEST, SO, IN THIS CASE, YOU'D SAW IT THE WAY I DID, AND "SNUG UP" A COUPLE OF WEDGES IN THE CUT—THEN YOU CAN CUT UP THE REST OF THE TREE, AND THE WEDGES WILL HOLD THE BUTT CUT OPEN···YOU CAN EASILY FINISH IT LATER!
EASY WHEN YOU KNOW HOW, EH!

LATER
YOU BOYS BETTER GO OVER TO THE CABIN AND GET YOUR SUPPER, BEFORE IT GETS COLD!
I HOPE IT'S NOT ONE OF YOUR *RELIGIOUS MEALS*, SIS··· YOU KNOW, *BURNT OFFERINGS*!
AW, HARV—THAT'S NOT NICE!
Bus Griffiths

YOU GUYS GO AHEAD--- I'LL BE ALONG!

THANKS FOR MAKING SUPPER, HONEY! HAVING YOU MAKE SUPPER---- IT'S KINDA LIKE WE WAS MARRIED--- I MEAN----
I KNOW---YOU'VE NEVER SAID A THING ABOUT US GETTING MARRIED BEFORE--- AND YOU'VE NEVER ASKED ME--- BUT I TOLD MY MAMA AND MY DADDY HOW I FEEL!

I'M SORRY--- I GUESS I NEVER THOUGHT MUCH ABOUT GETTING MARRIED---MAYBE BECAUSE I'VE GOT NOTHIN' TO OFFER YOU----
ALL I WANT IS YOU !!!

WELL, I SURE DON'T WANT YOU LIVIN' IN NO CAMP---AN' IT WOULDN'T BE ANY GOOD WITH ME IN CAMP, AND YOU AWAY UP HERE!
YOU COULD STAY UP HERE! DADDY LIKES YOU, AND WHEN HE KNOWS WE WANT TO GET MARRIED I KNOW HE'LL GIVE US SOME LAND! WE'D GET BY!

YOU MEAN QUIT LOGGIN'?
YOU COULD--- !?!

AW—NOW LISTEN, DEBBIE, AN' I WANT YOU TO UNDERSTAND---I'M A LOGGER! THAT'S WHAT I AM---IT'S WHAT I DO BEST! TO ME, LOGGING IS MORE A WAY OF LIFE THAN A JOB— I WOULDN'T BE THE SAME GUY IF I WASN'T LOGGING!
I DON'T REALLY THINK YOU'RE BEING FAIR!

A MAN DOES WHAT HE HAS TO— BUT THAT DOESN'T MEAN WE CAN'T WORK SOMETHIN' OUT, 'CAUSE I SURE DO LOVE YOU!
I'D BETTER GO! I'VE GOT TO HELP MAMA WITH SUPPER!
Bus Griffiths-

THE BIG TREES FELL, THE FRESHLY BUCKED LOGS ROLLED FREE, AND THE DAYS PASSED— THE FACE OF THE ROCKY DRAW BEGAN TO CHANGE DRAMATICALLY, TAKING ON AN APPEARANCE OF GREAT HAVOC!
THIS MORNING, THE MEN AWAKENED TO THE SOUND OF WIND IN THE TREETOPS, AND A DEFINITE CHILL IN THE AIR— AT DAYLIGHT, A SOFT MANTLE OF SNOW COVERS THE LANDSCAPE, HIDING THE RAW WOUNDS OF THE ROCKY DRAW!

WE WON'T BE FALLIN' ANY TIMBER IN THAT WIND!
NOPE— SHE LOOKS LIKE A HARBOUR DAY TO ME!
LISTEN TO THE OLD FISHERMAN!-- --IT'S ABOUT TIME WE HAD A DAY OFF! I'VE BEEN WORKIN' SO GODDAMNED HARD I'M GETTIN' MUSCLES IN MY BREATH!

IT WOULD SURE BE AN IDEAL DAY TO GET A BUCK!
SAY, THAT'S A GREAT IDEA! THE FOLKS COULD SURE USE SOME FRESH MEAT---AND A LITTLE VENISON WOULD BE PRETTY GOOD!

BREAKFAST, THIS MORNING, CONSISTS OF A BACON SANDWICH AND A CUP OF COFFEE, BUT EACH OF OUR FRIENDS WRAPS ANOTHER BACON SANDWICH IN WAX PAPER, AND TUCKS IT IN A MACKINAW POCKET----AND A SHORT TIME LATER, THE HUNT IS PLANNED IN THE CHILL, EARLY LIGHT OF A WINTER WONDERLAND—

THERE'S A RIDGE TWO MILES BACK— THAT'S AS FAR AS WE'LL GO— I'LL HUNT UP THIS SIDE, ONE OF YOU TAKE THE MIDDLE, AND THE OTHER CAN GO UP THE FAR SIDE— WE SHOULD GET A BUCK BEFORE WE REACH THE RIDGE!
OKAY, HARV!
I'LL TAKE THE FAR SIDE!

THE FRESHLY FALLEN SNOW SEEMS TO CHANGE THE TIMBERLAND INTO AN ENCHANTED FOREST, AND COVERS EVERYTHING WITH A BLANKET OF SILENCE— EVEN THE WIND IN THE TOPS OF THE TREES SOUNDS MUTED AND FAR AWAY— AL GETS THE STRANGE FEELING THAT IT WOULD BE CLOSE TO SACRILEGIOUS TO FIRE A SHOT AND BREAK THAT SILENCE— HE ALMOST HOPES THAT HE DOES NOT SEE A DEER!
BUS GRIFFITHS—

A CASUAL OBSERVER MIGHT SAY THAT AL WAS MORE INTERESTED IN STUDYING THE TIMBER THAN IN LOOKING FOR GAME, BUT HIS RESTLESS EYES ARE VERY BUSY SCANNING THE SNOW-COVERED LANDSCAPE, AND NOTE THE PASSING OF EVERY FOREST CREATURE—

TOPPING A SLIGHT RISE, AL AND RED COME TO AN ABRUPT HALT—
GOOD GOD! THERE'S HARV LAYIN' DOWN NEAR THAT BIG WINDFALL!
HARV! WHAT IN HELL HAPPENED?
AW···I SAW A BIG FOUR-POINT, AN' AS I SHOT, I SLIPPED AND TWISTED MY ANKLE!
DID YOU MISS THE BUCK?
I SHOULD'VE GOT HIM EASY! HE WAS JUST OVER THERE— COME FROM OVER YOUR WAY, AN' HE HAD A DOE WITH HIM— I HIT HIM, BUT HE TOOK OFF-- I DON'T THINK HE'S HURT BAD!
I'LL TAKE A LOOK!
H'M··· RIGHT FRONT LEG DRAGGING····· PROBABLY SHOT IN THE SHOULDER!
HARV, CAN YOU GET BACK TO YOUR PLACE WITH RED'S HELP?
SURE--SURE I CAN! BUT WHAT ARE YOU GONNA DO? YOU DON'T MEAN YOU'RE GOING AFTER THAT DEER, DO YOU? I DON'T THINK IT'S HURT TOO BAD!
DAMMIT, HARV--THE DEER IS HURT! IT LOOKS LIKE IT'S GOT A BUST SHOULDER— MAYBE A BROKEN LEG, TOO···SO, IF IT'S POSSIBLE, I'LL TRACK IT DOWN AND END ITS SUFFERIN'!
BUS Griffiths

I'LL WAIT ABOUT TWENTY MINUTES AND THEN TAKE UP THE TRAIL····· IF THE BUCK IS GONNA LAY DOWN THAT'LL GIVE HIM TIME TO STIFFEN UP!

OH, OH!····THE WIND HAS CHANGED···IT'S BLOWIN' FROM ME TO HIM···HE MIGHT KEEP GOIN' UNTIL HE DROPS!

HE'S BLEEDING PRETTY BAD, BUT SOMETIMES THEY SEEM ABLE TO GO A LONG WAY!

THE GROUND WAS GETTING MUCH STEEPER, AND AL COULD SEE ROCK OUTCROPPINGS BETWEEN THE TREES —
THIS MUST BE THE RIDGE HARV MENTIONED!

ONCE ACROSS THE RIDGE, THE GROUND PITCHES SHARPLY DOWNHILL—
LOOKS LIKE THAT BUCK IS HEADIN' RIGHT DOWN INTO THE NEXT VALLEY! ····BUT HE CAN'T GO MUCH FURTHER!

BOY! THERE'S SURE SOME FINE TIMBER IN HERE! JUST LOOK AT THOSE CEDARS!

WONDER WHO IN HELL OWNS THIS STUFF? I'D SURE LIKE···OH, OH! THOUGHT I SAW SOMETHIN' MOVE AWAY DOWN BELOW!
Bus Griffiths

A SHORT TIME LATER, AL HAS THE DEER DRESSED, THE LEGS CUT, AND THE ANIMAL TRUSSED UP FOR BACK PACKING—

HE WRESTLES THE CARCASS UP ONTO THE WINDFALL, THEN REMEMBERS THE BACON SANDWICH THAT HE HAD PUT IN HIS MACKINAW POCKET EARLIER THAT MORNING—

HE CLEANS HIS HANDS IN THE FRESHLY FALLEN SNOW—

IN THE FAILING LIGHT OF A WINTER AFTERNOON, A HEAVILY BURDENED FIGURE TRUDGES DOWNHILL THRU THE SNOW, DWARFED BY THE MAGNIFICENT STAND OF TIMBER THRU WHICH HE PASSES—

LADY LUCK WAS RIDING WITH AL! JUST AS DUSK SETTLED OVER THE LANDSCAPE, HE STEPPED ONTO THE END OF A NARROW STRIP OF LEVEL GROUND ON THE SIDEHILL, AND COULD FEEL HIS CAULKED BOOTS BITE INTO THE SNOW-COVERED CEDAR PUNCHEON OF AN OLD SKIDROAD—

SOMETIME LATER, AL NOTICES A LIGHTENING IN THE SKY AHEAD, AND THEN HE HEARS THE LOW MURMERING OF THE RESTLESS SEA— HE REALIZES HE IS APPROACHING THE BEACH, AND NEARING THE END OF HIS JOURNEY—

SUDDENLY AL BREAKS INTO THE OPEN, AND THE FULL EXPANSE OF THE BAY LIES BEFORE HIM— HE SEES THE LIGHTS OF THE BROWN'S HOMESTEAD OFF TO HIS RIGHT, AND OUT ON THE BAY ARE MORE LIGHTS—

Bus Griffiths-

QUIET, SHEP! AL, IS THAT YOU?
DEBBIE! WHAT ARE YOU DOING OUT IN THE COLD?

I WAS SO WORRIED ABOUT YOU--I IMAGINED ALL KINDS OF THINGS HAPPENING---NOW, YOU COME IN AND GET SOME SUPPER!
YOU SHOULDN'T WORRY ABOUT ME, BUT IT'S AWFUL NICE TO KNOW YOU DO! HOW'S HARV'S ANKLE?

DROP THE DEER THERE, AL! RED AN' I WILL HANG IT IN THE SHED! MY ANKLE IS OKAY NOW--AND THANKS FOR GETTING THE BUCK--I'M KIND OF ASHAMED OF MYSELF!
NO NEED TO FEEL THAT WAY, HARV!
SAY, AL—THE CAT'S HERE! THE TUG GOT IN JUST BEFORE DUSK---AN' ART'S UP, TOO! I HEARD HIM BELLOWING ORDERS AT THE SKIPPER---- TELLIN' HIM HOW, AND WHERE, TO ANCHOR THE TUG!

I HOPE IT'S NOT TOO GRIM WHEN ART COMES ASHORE! HE'LL HAVE ONE HELLUVA TIME KEEPIN' HIS LANGUAGE CLEAN IN FRONT OF THE WOMEN FOLK! I THINK PROFANITY IS HIS MOTHER TONGUE!
YEAH—IT MIGHT BE KIND OF PATHETIC! BUT THEN, HE MAY SURPRISE US, TOO!

YEAH—I HAD TO TRAIL HIM OVER THAT RIDGE AN' RIGHT INTO THE NEXT VALLEY—I WAS LUCKY AND HIT ONE OF THE OLD SKIDROADS, AN' CAME OUT THAT WAY!---- WHAT I'D LIKE TO KNOW IS--WHO OWNS THAT NICE TIMBER IN THERE?
BOY! ISN'T THAT SOMETHING? DAD KNOWS ALL ABOUT THAT STUFF!

THAT'S A BIG BLOCK OF TIMBER! IT WAS OWNED BY A FELLOW NAMED BEN HARDY, IN TORONTO-HE DIED LAST YEAR--- I SUPPOSE HIS WIDOW HAS IT NOW!
I'M GONNA GET ART TO LOOK AT IT! OUR BOSS, OLD JIM, HAS BEEN TALKING, FOR QUITE AWHILE, ABOUT GETTIN' MORE TIMBER! I THINK HE'LL BE REALLY INTERESTED IN *THAT* TIMBER!
Bus Griffiths

AL FIGURES THIS IS AN OPPORTUNE TIME TO TELL HIS BURLY FOREMAN, ART DONNEGAN, ABOUT THE MAGNIFICENT STAND OF TIMBER HE HAD SEEN THE PREVIOUS DAY, WHILE TRAILING THE WOUNDED BUCK DEER—

Bus Griffiths-

LATER WE FIND AL RICHARDS AND ART DONNEGAN DEEP IN THE BIG VALLEY, WHERE AL HAD SHOT THE BUCK DEER THE PREVIOUS DAY—

THIS IS BIG STUFF, AND THERE'S A LOT OF SCALE HERE, AN' IT'S A GOOD GRADE OF TIMBER!
IT LOOKS PRETTY SOUND! I HAVEN'T SEEN ONE TREE WITH *"ELEPHANT EARS" ON IT, AND THERE AREN'T TOO MANY *"SNAGS"!

THE GROUND IS PRETTY GOOD, TOO--AT LEAST WHAT WE'VE SEEN OF IT!
WELL, WE'VE BEEN WALKIN' THRU SOLID TIMBER FOR TWO HOURS! I'VE SEEN ENOUGH TO KNOW I'LL BE BACK WITH JIM--HE'S GOTTA SEE THIS!

JIM'S A CAGEY OLD ROOSTER AN' HE'LL SURE LOOK THINGS OVER BEFORE HE MAKES A MOVE--HE'LL CHECK OVER THE GROUND AND THE DRAINAGE, TO KNOW WHAT PROBLEMS HE'LL FACE IN ROAD BUILDING-HE'LL CHECK THE BEACH AND THE BAY TO KNOW WHAT KIND OF BOOMING GROUND HE'LL HAVE-IN FACT HE'LL CHECK EVERYTHIN'!

YOU KNOW, AL, THERE'S NOTHIN' LIKE A WALK THRU A STAND OF BIG TIMBER TO MAKE A MAN FEEL MIGHTY HUMBLE!
YEAH-IT SURE MAKES A GUY FEEL PRETTY INSIGNIFICANT, ALRIGHT!

YET US GUYS COME INTO A PLACE LIKE THIS AND WE CUT DOWN THE TREES, AN' WE HANDLE THE BIG LOGS LIKE THEY WAS NOTHIN'! EVERY DAY WE WORK, WE HANDLE LOGS AN' EQUIPMENT THAT WEIGH MANY TONS, AND WE COULD BE KILLED AT ANY TIME, AND WE DON'T THINK MUCH ABOUT IT! BUT WE'RE GOOD AT WHAT WE DO, AN' WE KNOW IT! I GUESS THAT'S WHAT MAKES US LOGGERS SUCH A PROUD, COCKY BUNCH OF BASTARDS!
I GUESS WE ARE A BREED APART, ALRIGHT!

Logging Terms

*"ELEPHANT EARS"- BRACKET FUNGI- ALSO CALLED "CONKS"— WHEN ATTACHED TO A TREE, IT SIGNIFIES THAT THE TREE IS "CONKY", OR ROTTEN

*"SNAGS"~ DEAD TREES WITH THE TOPS BROKEN OFF— A GREEN TREE WITH A BROKEN TOP IS OFTEN CALLED A "GREEN SNAG"~

Bus Griffiths—

WE MAY AS WELL HEAD BACK—I'VE SEEN ENOUGH! I'LL BE BACK WITH JIM IN A FEW DAYS!

SUDDENLY THE SILENCE OF THE FOREST IS BROKEN BY THE SOUND OF A FAR-AWAY BLAST—

IT RUMBLES THRU THE BIG VALLEY AND ECHOES OFF THE HEAVILY-TIMBERED SIDEHILLS LIKE A PEAL OF THUNDER—

NOW, *WHAT IN HELL* WAS *THAT*? SOUNDED LIKE SOMEBODY WAS DOIN' SOME BLASTING!
THERE'S A NARROW GULLEY WE FIGURED ON USING TO GET THE CAT DOWN INTO THE DRAW WHERE WE'RE LOGGING—I'LL BET RED AND HARV WERE BLASTING SOME OF THE ROCK TO WIDEN IT!

4
I WISH I HAD TIME TO GET UP AN' SEE WHAT KIND OF A ★"SHOW" YOU GOT, BUT MEBBE NEXT TIME—HOW'S THE TIMBER?
A LOT LIKE THIS STUFF, BUT NOT AS TALL—IT'S PRETTY GOOD TIMBER!

WHEN WE GET BACK I WANT TO MEET THIS "FATHER-IN-LAW" OF YOURS—I GOT A DEAL FOR HIM ON THEM LOGS YOU'RE TAKIN' OUT—ALSO I WANT TO LEARN EVERYTHIN' HE KNOWS ABOUT *THIS* TIMBER!
I'LL INTRODUCE YOU TO HIM, BUT HE AIN'T MY "FATHER-IN-LAW"!
6

WELL, FROM THE *LOOK OF THINGS*, AN' THE WAY YOU USED TO RUSH OVER TO THE OFFICE EVERY MAIL DAY, I'D SAY IT AIN'T GONNA BE TOO LONG UNTIL *HE IS* YOUR FATHER-IN-LAW! *AIN'T I RIGHT?*

WELL····I GUESS YOU COULD BE!

LOGGING TERMS

★"*SHOW*"—A PIECE OF COUNTRY TO BE LOGGED—A "GOOD SHOW" OR A "BAD SHOW"

★"*PIKE POLE*"—A BOOM-MAN'S MAIN TOOL—A LONG POLE WITH A PIKE-TIPPED END AS IN SKETCH—USED TO MOVE LOGS THAT ARE FLOATING FREE IN THE WATER—

★"*DOG LINE*"—A LONG LINE (CABLE) WITH AN EYE SPLICED IN EACH END—ONE EYE IS SPLICED INTO A "RING DOG" AND SEVERAL MORE DOGS ARE THREADED OVER THE LINE—USED PRIMARILY DRIVEN INTO FLOATING LOGS TO HOLD THEM TOGETHER—

LATER— UP IN THE WOODS—
YOU GOTTA BE KIDDING!
I'M NOT KIDDING, RED— IT'S THE TRUTH! ART HAD US BOTH FOOLED! HE'S A REAL LADIES' MAN! HE CHARMED BOTH DEBBIE AND HER MOM!
AFTER HE'D LEFT, MISSUS BROWN SAID, "WHAT A NICE MAN HE IS--- HE'S A LITTLE ROUGH AND READY, BUT HE'S A REAL GENTLEMAN"— SHE SAID WE WAS REAL LUCKY TO BE WORKIN' FOR SUCH A NICE MAN!
I DON'T BELIEVE IT! I JUST DON'T BELIEVE IT!
YOU ARE KIDDING---- AIN'T YOU?
NOPE—IT'S THE GOSPEL TRUTH, RED! NOW, YOU GUYS GET IN THE CLEAR!
TAKE 'EM AWAY, DOUG!
I STILL FIND IT HARD TO BELIEVE THAT STORY, AL!
WELL, YOU KNOW, RED— TRUTH IS STRANGER THAN FICTION!
BUS Griffiths

Logging Terms

*"IN THE BIGHT"—

GENERALLY MEANS TO BE INSIDE THE CURVE OF A DIVERTED LINE (CABLE), WHICH, WHEN UNDER STRAIN WILL STRAIGHTEN WITH DIRE RESULTS — IN THIS CASE IT MEANS TO GET IN ANY DANGEROUS SITUATION —

THAT WIND IS SURE FRESHENING, AND THOSE TOPS ARE REALLY STARTIN' TO MOVE ALONG!
THEY'RE MOVING NEARLY AS FAST AS I CAN WALK!
WELL, LOOK AT THAT! THEY'RE ROUNDING RIGHT INTO THE BAY... THE LOGS SHOULD DO THE SAME THING!
THAT MEANS THAT ANY DAY WHEN IT'S TOO ROUGH FOR HARV TO GO OUT OF THE BAY, WE CAN STILL DUMP LOGS INTO THE CHUCK!
AS LONG AS THE TIDE IS FLOODING, AND THE WIND IS BLOWING UP THE SOUND THE LOGS WILL COME RIGHT INTO THE BAY TO MEET HARV!
ALL HE'LL HAVE TO DO IS ROUND 'EM UP AND PUT 'EM INSIDE THE BOOMSTICKS... LIKE PUTTING IN THE CHICKENS!
Bus Griffiths—

THE DAYS PASS

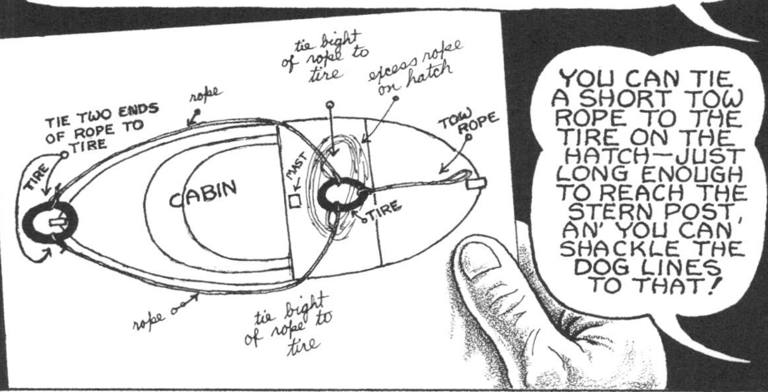

There was a great deal of excitement in the Brown household this morning—

Harv had left word that the first logs were due to "*hit the chuck" shortly before ten o'clock

We leave Debbie and her folks, as they hurry up the steep trail to a vantage point on the bluff, and turn to Al and Bruce, up in the woods—

Doug, the *"cat skinner" catches a big fir log with his bulldozer blade and pushes it towards the edge of the bluff- the log seems to teeter for a moment, then rushes down the short chute formed by the sloping rock and starts the long plunge to the water, far below

Logging Terms

*"Hit the Chuck"

To go into the salt water, the destination of all good logs—

◆ ◆ ◆

*"Cat Skinner"

Tractor operator

THE FIRST LOG HITS THE CHUCK WITH A "BOOM" LIKE A CANNON BLAST, AND AS IT DISAPPEARS BENEATH THE SURFACE OF THE WATER IN A CASCADE OF SPRAY, ANOTHER LOG STARTS THE LONG DROP —
RED AND HARV ARE COMING ALONG THE SHORELINE AS THE FIRST LOG HITS THE WATER!
WOW! LOOK AT THAT!
YEAH- AIN'T IT SOMETHIN?
BUT DON'T GO TOO CLOSE! THOSE LOGS GO UNDER A LONG WAY, AN' WHEN THEY COME UP THEY'RE LIKE A SPOUTING WHALE! IT'D BE A HELL OF A NOTE TO HAVE ONE OF THOSE BASTARDS COME UP THRU THE BOTTOM OF THE BOAT!
I'LL JUST IDLE HER INTO THE TIDE!
THAT TIDE IS REALLY STARTING TO PUSH THOSE LOGS— WE'LL LET THE FIRST THREE GO PAST US, AN' DOG THE FOURTH ONE— THEN WE'LL TURN AND PICK UP THE OTHER THREE ON THE WAY IN! SAY—HAVE YOU GOT A BRACE AND A SMALL BIT ABOARD ?
WHAT YOU WANT THIS FOR?- ARE YOU GONNA SCUTTLE MY BOAT?
I'M GONNA DRILL A HOLE IN THIS AXE HANDLE SO I CAN FASTEN A ROPE TO IT— THEN I'LL TIE THE ROPE TO ONE OF THE RINGS IN THE STERN··· IT'S AWFUL EASY TO LOSE AN AXE OVERBOARD WHEN YOU'RE DRIVING A DOG INTO A LOG!
Bus Griffiths

HAVE YOU GOT ANY HAYWIRE ABOARD THIS TUB ?
WHAT KIND OF AN OUTFIT DO YOU THINK I'M RUNNIN'? OF COURSE I'VE GOT SOME HAYWIRE ABOARD!

DO YOU SEE HOW THE EYE OF THE LINE IS SPLICED INTO THE RING ON THAT DOG?.... AND THE REST OF THE DOGS ARE THREADED OVER THE LINE ?
YEP!

WELL, IF YOU STARTED TOWING, ALL THESE DOGS WOULD SLIDE DOWN THE LINE, AND WHEN YOU WANTED TO PICK UP ANOTHER LOG YOU'D HAVE TO PULL THE LINE IN TO GET A DOG— BUT--
AH! I SAVVY NOW!—I'LL GET THE HAYWIRE!

IF YOU TWIST A PIECE OF HAYWIRE ON THE SHACKLE AND BEND A HOOK ON EACH END, YOU CAN HOLD THE DOGS UP AT THIS END OF THE DOG LINES, AND GET 'EM WHEN YOU NEED 'EM!
WONDERFUL STUFF, THIS HAYWIRE! I DON'T KNOW WHAT US FISHERMEN, AN' YOU LOGGERS WOULD DO WITHOUT IT!

A SHORT TIME LATER, THE FIRST TOW OF LOGS HEADS UP THE SOUND—
Bus Griffiths

THE WEATHER HOLDS GOOD, THE MEN WORK HARD, AND THE TIME PASSES--- UNTIL ONE DAY---
C'MON, C'MON! THROW THAT CHOKER AROUND THAT LOG, AN' LET'S SEE A BIT MORE ACTION AROUND HERE! MY OLD GRANDMOTHER CAN MOVE FASTER THAN THAT--- AN' SHE'S NINETY YEARS OLD!
ART!

WHEN DID YOU GET HERE? I DIDN'T KNOW YOU WAS EVEN UP HERE!
HELL! WE COME UP THREE DAYS AGO—WE'VE BEEN CAMPIN' UP IN THE TIMBER! ME, OLD JIM, AN' LEN ARMITAGE--- HE'S HEAD FOREST RANGER FOR THIS DISTRICT!

THEY WENT OUT THE OTHER WAY—JIM WANTED TO SEE YOUR "FATHER-IN-LAW" AN' LOOK OVER THE BEACH—I FOLLOWED YOUR OLD TRACKS IN THE SNOW AN' GOT UP HERE--- YOU'VE GOT A PRETTY GOOD SHOW HERE!
YEAH—WE'RE TRYIN' NOT TO WASTE ANY OF THE GOOD STUFF!

WE PASSED RED AN' YOUR "BROTHER-IN-LAW" ON THE WAY IN! THEY SURE ARE PUTTIN' A LOT OF NICE WOOD IN THAT BOOM! GOT MUCH LEFT?
JOE DON'T WANT ANY MORE TIMBER CUT--- WE'LL JUST TAKE OUT WHAT'S DOWN—BUT EVEN IF THE WEATHER HOLDS, WE WON'T GET THRU BY CHRISTMAS!

WELL, JUST HANG TOUGH—JIM MADE A DEAL WITH YOUR "FATHER-IN-LAW" AN' WE'RE BUYIN' THE LOGS! YOU ARE WORKIN' FOR US AGAIN!
OH-H!--- WHAT DID JIM THINK OF THAT OTHER TIMBER?

WELL, WE SPENT THREE DAYS LOOKIN' IT OVER WITH THE RANGER! THAT'S A BIG BLOCK OF TIMBER, AN' JIM WANTS IT! HE'LL BE PHONIN' TORONTO WHEN WE GET BACK TO CAMP--- HIS LAWYER IS ALREADY THERE!--SOMETHING BIG MIGHT COME OUTA THIS!

I BETTER BEAT IT AN' LET YOU GET BACK TO WORK—WE'RE LOSING MONEY WITH YOU STANDIN' AROUND! PROBABLY SEE YOU AGAIN IN A FEW DAYS!
YEAH, SURE!
DAMMIT!---AN' ALL THE TIME I FIGURED WE'D BE LOGGIN' ON OUR OWN!
BUS Griffiths—

DURING THE PAST FEW DAYS, AL'S HEART HASN'T BEEN IN HIS WORK— ART DONNEGAN'S WORDS OF A FEW DAYS AGO, "YOU'RE WORKIN' FOR US AGAIN" HAVE TURNED WHAT HAD BEEN AN EXCITING CHALLENGE INTO JUST ANOTHER JOB— AND TO FURTHER WORSEN MATTERS, THE WEATHER HAS TURNED SOUR TO MATCH AL'S MOOD— RAIN HAS FALLEN STEADILY FOR THE PAST TWO DAYS & THE CAT ROAD IS A QUAGMIRE OF MUD— AL HAS SENT YOUNG BRUCE HOME, AS THE LAD HAD NO PROPER RAIN CLOTHES— WORKING BEHIND A CAT IN WET WEATHER IS A TOUGH, DIRTY JOB & A DISGRUNTLED AL RICHARDS, DRIPPING MUD & WATER, HIS CAULKED BOOTS UNRECOGNIZABLE LUMPS OF MUD, DRAGS UNWILLING FEET BEHIND THE CAT & A TURN OF LOGS—

I GUESS IT REALLY DON'T MAKE MUCH DIFFERENCE---- I'M STILL DOIN' THE SAME THING AS I WAS BEFORE--- BUT THE MISERABLE BASTARD DIDN'T NEED TO PUT IT THE WAY HE DID!

I KNOW THEY OWN THE EQUIPMENT, BUT WE'RE DOIN' THE LOGGIN'!---- HE COULDA SAID, "WHEN YOU FINISH LOGGIN' THAT TIMBER, WE'D LIKE TO BUY THE LOGS!"--- OR SOMETHING LIKE THAT!

THAT'S YOUNG BRUCE! HOPE NOTHING'S WRONG AT THE HOUSE! WHAT'S UP, BRUCE?
BUS GRIFFITHS

THOSE TWO BIG MEN CAME UP IN A SPEEDBOAT--- I THINK YOU CALLED THEM "ART" AND "JIM"— THEY SAID FOR YOU TO QUIT WORK AND COME DOWN---- IT'S VERY IMPORTANT! THEY'RE COMING UP TO MEET YOU--- BUT THEY TALKED FUNNY, AND THEY LAUGHED A LOT!
THANKS, BRUCE! NOW YOU SCOOT BEFORE YOU GET SOAKED!

A SHORT TIME LATER—

THERE THEY ARE! I WONDER WHAT'S UP? I DOUBT THEY'D MAKE A SPECIAL TRIP JUST TO TELL ME THEY GOT THE TIMBER! SAY---- THEY LOOK LIKE THEY'RE HALF CUT!
HI! YOU GUYS WANTED TO SEE ME?
HELLO LAD!
WHO IS THIS DIRTY MUDDY WAYFARER? GIMME THE BOTTLE, JIM--- THE POOR BASTARD LOOKS LIKE HE COULD USE A DRINK!

GRAB A SEAT AN' TAKE A GOOD SLUG, AL! YOU LOOK KIND OF MEAN AN' MISERABLE!
I FEEL MEAN AN' MISERABLE, TOO!
WELL, LAD– WE'VE GOT THE TIMBER! WE HOPE TO GET THINGS STARTED IN THE SPRING!
GOOD! ...BUT YOU DIDN'T MAKE THE TRIP UP HERE, JUST TO TELL ME THAT!
NOPE!...AN' WE WOULDN'T HAVE COME ATALL IF WE'D KNOWN YOU WAS GONNA BE SO MASACHI! WHY CAN'T YOU BE LIKE ME?
WELL, LAD– WE'VE GOT A PROPOSITION TO OFFER YOU!
HOW WOULD YOU LIKE TO GET THIS NEW CAMP GOIN'...TAKE IT OVER...AND RUN IT?
WHO, ME !?!
WHY NOT YOU? YOU'VE DONE A GOOD JOB ON EVERYTHING WE'VE EVER ASKED YOU TO DO!
YOU BEST GIVE ME ANOTHER GO AT THAT BOTTLE...I THINK I'M HEARIN' THINGS!
YOUR HEARIN' IS OKAY! WE'VE DECIDED...IT'S YOU AN' RED! YOU'LL BE IN CHARGE....THE SUPER!... SOUNDS PRETTY GOOD, EH? RED CAN BE YOUR PUSH!
AS LONG AS THINGS ARE GOIN' GOOD, THERE WILL BE NO INTERFERENCE FROM US! WHAT YA SAY?
MAN! HOW COULD I TURN IT DOWN? SURE I'LL TAKE IT! THANKS FOR GIVIN' ME THE CHANCE !!!
HELL...WE FIGURED YOUR MAIN INTEREST IS UP HERE, AND WE WOULDN'T GET MUCH WORK OUTA YOU IF WE TOOK YOU BACK TO CAMP...YOU'D BE MOONIN' ABOUT THIS PLACE ALL THE TIME!
HAVE ANOTHER DRINK, LAD...AN' THEN WE'VE GOT TO GET BACK TO CAMP...AN' ABOUT YOUR SALARY.... WE'LL MAKE IT REALLY WORTH-WHILE!
WELL– NOW YOU'VE GOT NO BLOODY EXCUSE FOR NOT MAKIN' THAT POOR OL' MAN YOUR "FATHER-IN-LAW"!
I'D BETTER GO UP TO THE CABIN AN' GET CLEANED UP... AN' SOBER UP! I CAN'T LET HER SEE ME LIKE THIS!
BUS GRIFFITHS

IT WAS EARLY THE FOLLOWING MORNING WHEN AL LEFT THE CABIN AND HEADED FOR THE BROWN HOMESTEAD— THE RAIN HAD STOPPED, THE SKY WAS CLEARING, AND IT SHOWED PROMISE OF TURNING INTO A BRIGHT SUNNY MORNING—

THERE WOULD BE NO WORK TO-DAY! THIS, AL HAD DECIDED, WOULD BE THE *BIG DAY!*

HE TIMED HIS ARRIVAL AT THE HOUSE DURING THE PERIOD WHEN DEBBIE, THE BROWN'S LOVELY, YOUNG DAUGHTER, WOULD BE DOING THE MORNING MILKING—

AL LOST NO TIME IN TELLING THE BROWN FAMILY, AND HIS SIDEKICK, RED HARRIS, THE NEWS ABOUT THE NEW CAMP!

A SHORT TIME LATER, DEBBIE RETURNS FROM THE BARN WITH THE MORNING'S MILKING

Bus Griffiths

AND AS THE SUN SINKS BEHIND THE SNOW-CLAD MOUNTAINS, WE LEAVE OUR TWO YOUNG LOVERS AS THEY SIT ATOP THE ROCKY BLUFF, OVERLOOKING THE SEA, THEIR HEARTS FULL OF LOVE AS THEY FACE THE BRIGHT FUTURE—

DEBBIE, WITH THE MAN SHE LOVES AT HER SIDE AND HER HEAD FULL OF THE DREAMS THAT YOUNG GIRLS DREAM—

AND AL, WITH HIS DREAM-GIRL, A NEW RESPONSIBLE JOB, AND A FOREST FULL OF TREES TO LOG····· ISN'T THAT WHAT ALL GOOD LOGGERS DREAM ABOUT?